Procedural Content Generation with Unreal Engine 5

Harness the PCG framework to take your environment design
and art skills to the next level

Paul Martin Eliasz

Procedural Content Generation with Unreal Engine 5

Group Product Manager: Rohit Rajkumar

Publishing Product Manager: Neha Pande

Book Project Manager: Srinidhi Ram

Senior Content Development Editor: Feza Shaikh

Technical Editor: K Bimala Singha

Copy Editor: Safi s Editing

Indexer: Rekha Nair

Production Designer: Aparna Bhagat

Marketing Coordinators: Nivedita Pandey

First published: November 2024
Production reference: 2060126

Published by Packt Publishing Ltd.
Grosvenor House
11 St Paul's Square
Birmingham
B3 1RB, UK.

ISBN 978-1-80107-446-9
www.packtpub.com

This book has been a long time in the making, and I am deeply grateful to my loving family and friends for their unwavering support and encouragement throughout this relentless journey. Their belief in me has been a constant source of motivation, inspiring me to strive to be a role model. A special thanks to my parents, whose inspiration and guidance have propelled me to pursue great things and persist toward a noble goal, no matter the challenges.

– Paul Martin Eliasz

Contributors

About the author

Paul Martin Eliasz is an experienced senior technical artist, educator, and consultant, with a 12-year background in 3D CGI and real-time game engine expertise. He leads his own studio, specializing in game development and 3D gamified applications tailored for web streaming (pixel streaming), VR, virtual production, XR, and various computer platforms. Throughout his career, he has contributed to multiple projects for esteemed clients such as Netflix and Bentley and has provided consulting services to various companies and studios, including Outernet Global and PureWeb. Currently, he is actively involved in mentoring and educating for CG Spectrum, Mastered, CAVE Academy, Symetri, Bentley, Dyson, and Epic Games EMEA. Additionally, he occasionally conducts educational sessions in universities across Europe and Asia and creates customized educational content for aspiring Unreal artists and Unreal tech artists who want to work in the gaming, animation, automotive, and virtual production space.

About the reviewers

Florian Becquereau is an Unreal developer on a quest to become a full-on Unreal expert. After wanting to be a game developer and working on his own game engine, he ended up working in QA and localization for 6 years in some large video game companies. For the last 3 years, he has put all his energy into Unreal Engine and Unreal Editor for Fortnite. He enjoys teaching and empowering people, breaking down overwhelming topics into simple digestible bites.

Shekar Vallakeerthy has been working in Pixelloid Studios (India) for the past 15 years as a senior 3D generalist and senior faculty professional. He completed the Unreal World Building Fellowship in 2023. He's also a member of the Epic Games and Autodesk Developer Community with a strong knowledge of the visual effects and gaming pipeline from concept to final output. He's been certified by Autodesk Maya and Unreal Engine, supplementing his professional experience with a solid academic background. He's engaged in Maya and Unreal webinars to inform his classes of the latest trends and technological developments for young aspiring students. Over the years, he has trained more than 5,000 students in India.

I would like to extend my heartfelt thanks to Raj Potula, Nayeem Akthar, Bala Shankar Bonthu, and Chadalawada Chandrasekhar Rao for their exceptional help in my VFX journey. Their creativity, technical expertise, and dedication have played a crucial role in achieving the high-quality results that have set new standards for our team.

Sulbha Jadhav is a UE-5 Authorized Instructor and serves as an adjunct professor in the immersive media design department at a reputed university. She excels in Blueprint scripting, level design, and 3D modules, including AR/VR technicalities. With extensive experience, she plays a pivotal role in imparting knowledge at universities and conducting impactful UE-5 training bootcamps. As a 3D QA professional, she specializes in testing vital synthetic data for computer vision, advancing visual AI. Committed to industry advancements, Sulbha contributes significantly to 3D design and emerging technologies. She also reviewed *Architectural Visualization in Unreal Engine 5* by Ludovico Palmeri.

As a technical reviewer for a UE-5 book, I dedicated significant time to researching and testing data in this ever-evolving field. I want to express my deepest gratitude to my daughter and close friends, whose unwavering support made this journey possible. Their belief in me and understanding during challenging moments were invaluable. Thank you for being my pillar of strength throughout this endeavor.

Zubaida Nila is a virtual production engineer, Unreal technical artist, and an Unreal Engine Authorized Instructor Partner. She combines her technical expertise with academic research in In-Camera Visual Effects (ICVFX) and supervises high-profile TVC and film post-production projects, including work for Netflix Asia, PUBG, and Petronas. With a strong focus on advancing virtual production and visual effects, Zubaida plans to author a technical book in the near future. She is also a member of the Technical Advisory Board for Unreal Engine professionals at Packt.

Table of Contents

6

Building a PCG Graph with Landscape Materials   211

7

Let's Build a Building Using the PCG Spline Controller   253

8

Building Biomes: Mastering PCG for Rich Environments   315

9

Creating Dynamic Animated Crowds with PCG   351

Part 3: Mastering Optimization and Elevating Your PCG Environments

10

Exploring Optimization, Debugging, and Performance Tools   391

11

Cheat Sheets, Extra Tips, and Shortcuts   425

Preface

Hello there! *Procedural Content Generation with Unreal Engine 5* is designed to help 3D artists or game designers with fundamental technical knowledge in Unreal Engine 5, helping them gain skills in developing a toolset for game environments, level design and even dynami content systems. If you possess a basic understanding of blueprints and have experience in creating simple games or designing environments, this book will enable you to maximize a PCG framework benefits and grow your expertise.

Who this book is for

Unreal artists who are already experienced with Unreal Engine will discover a fresh perspective on world-building with PCG tools through this book. It introduces a novel approach to constructing procedurally generated environments by leveraging the PCG framework and other elements, emphasizing the strategic use of specific nodes in a blueprint-node setup. Through this guide, artists can explore innovative techniques to improve their skills in Unreal Engine and create dynamic, procedurally generated worlds.

The three primary personas targeted by this content are as follows:

- **Level and environment artists**: Artists aiming to build open-world environments for their games, animation, and virtual production
- **Aspiring tech artists**: Individuals looking to develop tools for environments, helping other artists achieve results more efficiently
- **Game developers**: Developers wanting to understand the PCG framework and design their own games and interactive applications

What this book covers

Chapter 1, Introducing Procedural Content Generation, explores the theory and practical applications of procedural generation. Before diving into the workflow, we will explain what procedural content generation is, its origins, and best practices for real-world game development and application projects.

Chapter 2, Let's Create Our First Forest!, shows how to create a procedurally generated forest for gaming or animation. You'll design a forest with realistic trees and vegetation, allowing flexibility to modify foliage meshes. We'll cover integrating and merging node structures in a single PCG graph, combining trees, and adding varied grass, bushes, and stones to the landscape.

Chapter 3, Building Blueprints with PCG Component, looks at creating and integrating a custom PCG blueprint with the PCG component. You'll use the Point loop to spawn static meshes, combining patterns for unique arrangements. The tutorial covers crafting independent blueprints for unique PCG components in Unreal Editor and developing your own algorithms within the PCG blueprint for hands-on experimentation.

Chapter 4, Developing and Optimizing the Procedural Content Generation Tool, focuses on refining your PCG Actor blueprint by enhancing the PCG graph with additional nodes. You'll introduce new variables within the Actor blueprint, expanding its functionality and versatility. You'll also learn how to select variables that enable direct communication with blueprints, allowing for more interactive modifications within Unreal Engine.

Chapter 5, Building Spline Controllers with PCG Graph, shows how to create a forest with a river using the Unreal Engine water plugin, manipulate rivers, and combine various PCG graphs into a single one. You'll integrate this into an Actor Blueprint, enhancing your worldbuilding skills in Unreal Engine. Topics include creating foliage, the Forest Actor Blueprint, PCG graphs, and making adjustments for better outcomes.

Chapter 6, Building a PCG Graph with Landscape Materials, explores creating a tool for randomly placing vegetation and generating foliage around structures and rivers. It introduces using landscape materials to guide static mesh placement and integrate spline integrate spline component for a dynamic PCG system that adapts to the landscape.

Chapter 7, Let's Build a Building Using the PCG Spline Controller, shows how to create buildings using PCG, including walls with window gaps and multi-floor structures. We'll explore procedural methods for urban building creation and develop algorithms in a PCG graph to control building features. Tasks will focus on using simple cubes to build walls, with all work integrated within the PCG graph and Actor Blueprint.

Chapter 8, Building Biomes: Mastering PCG for Rich Environments, covers using the PCG Biome plugin to create diverse and dynamic biomes and design varied landscapes with procedural techniques. You'll learn how to generate terrain and foliage using datasets and texture information, customizing elements beyond the PCG graph. Sections will focus on foliage placement and new PCG techniques to build realistic, immersive biomes.

Chapter 9, Creating Dynamic Animated Crowds with PCG, shows how to spawn and animate crowds using the PCG plugin. You'll set up Actor Blueprints, integrate the PCG component, and create a PCG graph for character setup. Using Mixamo models, you'll spawn characters in a spline loop and enable them to walk independently with basic AI. This chapter provides hands-on experience with advanced PCG features for realistic crowd simulation.

Chapter 10, Exploring Optimization, Debugging, and Performance Tools, will guide you in optimizing the PCG tool for your project, focusing on enhancing fluidity and performance. You'll learn how to launch your project using PCG and create impressive environments without performance issues. We'll cover examples and methods to boost project efficiency and manage assets for procedural generation.

Chapter 11, Cheat Sheets, Extra Tips, and Shortcuts, provides tips on understanding PCG nodes and their functions using visual examples. We'll explore node formations, use sample meshes, and enable debug mode to illustrate concepts. You'll gain insights into building various PCG structures and compare the applications of PCG volumes and splines. These elements are crucial for precise control over asset placement and behavior in procedural generation.

To get the most out of this book

Software/hardware covered in the book	OS requirements
Unreal Engine 5 (Minimum 5.4)	Windows
GPU (NVIDIA RTX, AMD Radeon RX with 8+ GB VRAM), and at least 16 GB RAM. For more info, visit the official Epic Games website: `https://dev.epicgames.com/documentation/en-us/unreal-engine/hardware-and-software-specifications-for-unreal-engine?application_version=5.4`	

If you are using the digital version of this book, we advise you to type the code yourself or access the code via the GitHub repository (link available in the next section). Doing so will help you avoid any potential errors related to the copying and pasting of code.

Download the example code files

You can download the example code files for this book from GitHub at `https://github.com/PacktPublishing/Procedural-Content-Generation-with-Unreal-Engine-5`. In case there's an update to the code, it will be updated on the existing GitHub repository.

We also have other code bundles from our rich catalog of books and videos available at `https://github.com/PacktPublishing/`. Check them out!

Code in Action

Code in Action videos for this book can be viewed at `https://packt.link/0ZQQD`.

Conventions used

There are a number of text conventions used throughout this book.

`Code in text`: Indicates code words in text, database table names, folder names, filenames, file extensions, pathnames, dummy URLs, user input, and Twitter handles. Here is an example: "Last but not least is the tiniest asset in the tree megascan package, `SM_EuropeanBeech_Seedling_04_PP`."

Bold: Indicates a new term, an important word, or words that you see onscreen. For example, words in menus or dialog boxes appear in the text like this. Here is an example: "Make sure to change your **Import** settings to **Nanite** from the drop-down box as this will save you lots of time later when we get to the optimization of our assets! "

> **Tips or important notes**
> Appear like this.

Get in touch

Feedback from our readers is always welcome.

General feedback: If you have questions about any aspect of this book, mention the book title in the subject of your message and email us at customercare@packtpub.com.

Errata: Although we have taken every care to ensure the accuracy of our content, mistakes do happen. If you have found a mistake in this book, we would be grateful if you would report this to us. Please visit www.packtpub.com/support/errata, selecting your book, clicking on the Errata Submission Form link, and entering the details.

Piracy: If you come across any illegal copies of our works in any form on the Internet, we would be grateful if you would provide us with the location address or website name. Please contact us at copyright@packt.com with a link to the material.

If you are interested in becoming an author: If there is a topic that you have expertise in and you are interested in either writing or contributing to a book, please visit authors.packtpub.com.

Reviews

Please leave a review. Once you have read and used this book, why not leave a review on the site that you purchased it from? Potential readers can then see and use your unbiased opinion to make purchase decisions, we at Packt can understand what you think about our products, and our authors can see your feedback on their book. Thank you!

For more information about Packt, please visit packt.com.

Subscribe to Game Dev Assembly Newsletter!

We are excited to introduce Game Dev Assembly, our brand-new newsletter dedicated to everything game development. Whether you're a programmer, designer, artist, animator, or studio lead, you'll get exclusive insights, industry trends, and expert tips to help you build better games and grow your skills. Sign up today and become part of a growing community of creators, innovators, and game changers. `https://packt.link/gamedev-newsletter`

Scan the QR code to join instantly!

Share Your Thoughts

Once you've read *Procedural Content Generation with Unreal Engine 5*, we'd love to hear your thoughts! Scan the QR code below to go straight to the Amazon review page for this book and share your feedback.

`https://packt.link/r/1801074461`

Your review is important to us and the tech community and will help us make sure we're delivering excellent quality content.

Free Benefits with Your Book

This book comes with free benefits to support your learning. Activate them now for instant access (see the "*How to Unlock*" section for instructions).

Here's a quick overview of what you can instantly unlock with your purchase:

PDF and ePub Copies   **Next-Gen Web-Based Reader**

Access a DRM-free PDF copy of this book to read anywhere, on any device.

Use a DRM-free ePub version with your favorite e-reader.

Multi-device progress sync: Pick up where you left off, on any device.

Highlighting and notetaking: Capture ideas and turn reading into lasting knowledge.

Bookmarking: Save and revisit key sections whenever you need them.

Dark mode: Reduce eye strain by switching to dark or sepia themes

How to Unlock

Scan the QR code (or go to `packtpub.com/unlock`). Search for this book by name, confirm the edition, and then follow the steps on the page.

UNLOCK NOW

Note: Keep your invoice handy. Purchases made directly from Packt don't require one

Part 1:
The Fundamentals of
Procedural Content Generation

In this part, you will dig into the fundamental theory and practical applications of procedural generation, starting with your first **Procedural Content Generation (PCG)** examples. You will gain an in-depth understanding of PCG and learn how to create your first environment using this framework. Additionally, we will cover best practices for building supplementary components to enhance the environment's capabilities and ensure continuous improvement.

This part has the following chapters:

- *Chapter 1, Introducing Procedural Content Generation*
- *Chapter 2, Let's Create Our First Forest!*
- *Chapter 3, Building Blueprints with PCG Component*
- *Chapter 4, Developing and Optimizing the Procedural Content Generation Tool*

1

Introducing Procedural Content Generation

Welcome to this thrilling journey into the realm of procedural content generation within Unreal Engine 5!

I can sense your excitement as you prepare to explore the realm of 3D content and its procedural generation capabilities in game projects and real-time visualizations for VFX and film sets. Before we dive into practical applications, we should gain a deeper understanding of procedural generation and its various use cases.

In this chapter, I'll cover the fundamental concepts of procedural content generation, followed by an exploration of the disparities between traditional procedural generation methods and the enhanced capabilities introduced with the **Procedural Content Generation** (**PCG**) plugin tool in Unreal Engine 5.

This exploration will help you identify the most suitable approach for your needs.

I'll then guide you through the practical implementation of the PCG tool in your project, demonstrating how it can be tailored to your specific tasks. Following this, you'll have the opportunity to explore using PCG and learn how to align it within your project in Unreal Engine 5.

> **Note**
>
> A quick heads-up for users of the latest Unreal Engine 5, particularly version 5.4: be aware that certain nodes have been deprecated and substituted with the new PCG nodes. Due to these changes, we'll be utilizing the 5.4 version of Unreal Engine 5 for our discussions. Further clarification on this topic will be provided in the upcoming chapters.

In this chapter, we're going to cover the following main topics:

- What is PCG?
- Where and how to use PCG tools
- Understanding the PCG framework
- Building your first PCG tool

By the end of this chapter, you will have a solid understanding of PCG and how to use it. You will learn the basics of PCG, how it is used, and how it compares to traditional methods of procedural content generation. This topic will be very useful to cover the most information.

Free Benefits with Your Book

Your purchase includes a free PDF copy of this book along with other exclusive benefits. Check the *Free Benefits with Your Book* section in the Preface to unlock them instantly and maximize your learning experience.

Technical requirements

You will need a good computer that can run a PCG framework plugin and to install the following hardware and software to complete this chapter:

- Use the latest operating system, either Windows 10 or Windows 11

  Please note that this book is intended exclusively for Windows users and does not offer support for macOS.

- A GPU (an NVIDIA RTX or AMD Radeon RX with 8+ GB VRAM) and at least 16 GB RAM. For more information, visit the official Epic Games website: `https://dev.epicgames.com/documentation/en-us/unreal-engine/hardware-and-software-specifications-for-unreal-engine?application_version=5.4`.

- You will need to install the latest version of Unreal Engine 5.4 because this book is based on that version. For detailed instructions on how to install Unreal Engine 5.4, please visit `https://dev.epicgames.com/documentation/en-us/unreal-engine/installing-unreal-engine`. The reason for this requirement is that the latest PCG framework includes new nodes, and some nodes present in version 5.2 have been removed.

The code files for the chapter are placed at `https://github.com/PacktPublishing/Procedural-Content-Generation-with-Unreal-Engine-5`

The code in action video for the chapter can be found at `https://packt.link/GClp8`

What is PCG?

In computing, **procedural generation** is a technique where data is automatically generated through the application of an algorithm as opposed to being manually crafted. This method typically involves a combination of human-designed assets and algorithms that use computer-generated random numbers and computational capabilities.

Over the past two decades, game content, encompassing terrain, characters, items, and every element in game production, has typically been crafted using two primary approaches.

The first involves manual content creation, where artists produce 3D assets and strategically place them within the game environment. Here is an example of some trees created using the standard foliage tool in Unreal Engine:

Figure 1.1 – Unreal foliage tool example

The following image is an example of PCG enabling the creation of environments within the level:

Figure 1.2 – An example of an Unreal environment built with the PCG framework

Here is the PCG graph that illustrates the PCG node structure used to generate the forest shown in *Figure 1.2*:

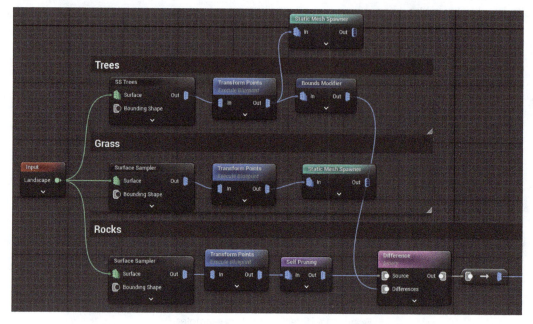

Figure 1.3 – A PCG graph node setup that creates a complete forest environment

The second approach is PCG, where algorithms autonomously generate content either prior to the initiation of a level or dynamically throughout the runtime of the game. This is highly beneficial when working on gameplay involving open-world levels with massive environments. It significantly improves the ability to stream levels if the environment contains a large volume of high-quality foliage instances. The benefits of this method include reduced file sizes, a broader spectrum of content, and the ability to establish more efficient structures for building levels and environments.

Figure 1.4 – An example of an Unreal environment built with the PCG framework

In the next section, we will explore the different types of procedural generation in the context of the PCG tool in Unreal Engine 5.

Where and how to use PCG tools

The game industry provides a distinct environment for developers and artists to innovate, especially in art and commercials. In this space, they can navigate the complex challenges of creating and managing a vast array of technically and creatively demanding 3D assets scattered throughout a level.

However, this method has evolved in recent years, particularly within the domain of real-time game engines, where capabilities have experienced exponential growth.

It's worth noting that, depending on the specific focus area, the use of PCG tools is not limited to game development alone. In the movie industry, in **visual effects** (**VFX**), for instance, PCG tools find valuable applications. The film industry, which has been accustomed to PCG for an extended period, has used software such as Houdini, which has been prevalent in this field for many years. Houdini is renowned for its procedural generation capabilities, enabling artists to create complex environments using node-based workflows, and remains a standard tool in the VFX industry today.

Unreal Engine excels at handling PCG, allowing the automated creation of complex and dynamic environments. This approach enhances visual diversity and reduces manual workload, enabling the quick generation of large, detailed landscapes.

The PCG tools in Unreal Engine, available as a plugin, streamline this process further. With intuitive, node-based workflows, artists and level designers can easily set rules for asset placement, automating tasks such as populating forests or laying out cityscapes. This empowers creators to focus on creativity, resulting in immersive, richly detailed environments and a more efficient development pipeline.

Introducing the Unreal Engine PCG plugin

The PCG plugin operates by evaluating a component known as **Hierarchical Instanced Static Meshes (HISM)**. This HISM component enables the generation of an extensive array of static meshes across a surface. PCG utilizes this information to project the spawn positions of instances onto the designated surface, such as a landscape, in the form of **PCG volumes**. Alternatively, **Instanced Static Meshes (ISM)** can be used, but HISM is the default and preferred component due to its much more convenient ability to render multiple static meshes with a single draw call without any loss of performance.

In summary, HISM and ISM are powerful components for optimizing the rendering of large numbers of instances, which is often the case in procedural generation systems. They allow you to efficiently distribute and render instances across surfaces such as landscapes while providing flexibility for dynamic adjustments.

However, they have some key differences:

- **Regular ISM**

    Each instance in an ISM is treated independently and can have its own transform, material properties, and so on. ISM components are suitable when you have a large number of identical or nearly identical static meshes that don't share common characteristics, and you want each instance to be treated individually.

    While ISM components provide a basic **Level of Detail (LOD)** support, each instance can switch between different LOD meshes based on distance from the camera. This helps in maintaining performance by reducing the detail of distant instances.

    Unlike HISM, ISM does not support hierarchical culling. However, it still benefits from reduced draw calls and improved performance by using instanced rendering and basic LOD transitions.

- **HISM**

    HISM components are designed for situations where you have a large number of instances that share common characteristics and can be grouped hierarchically:

    - Instances in a HISM component share the same material properties and other characteristics, and they are organized in a hierarchical structure. This means that you can efficiently cull entire branches of the hierarchy if they are not visible, reducing the rendering cost.

- HISM components are particularly useful for things such as foliage systems, where you have many instances of the same type of mesh (e.g., trees, grass, bushes) that can be grouped together.

- HISM components allow dynamic LOD management based on the distance of the instances from the camera. Closer instances are rendered with higher detail, while farther instances use lower-detail meshes, reducing the rendering workload.

- HISM components support hierarchical culling, which means entire groups of instances can be culled (not rendered) if they are outside the camera view or beyond a certain distance, further optimizing performance

In the next part, we will explore and compare ISM and the PCG tool, focusing on their respective performances and differences.

Exploring ISM and HISM

Here are some examples of the use of ISM and HISM. We have used the SM_Rock asset from the Starter Content folder:

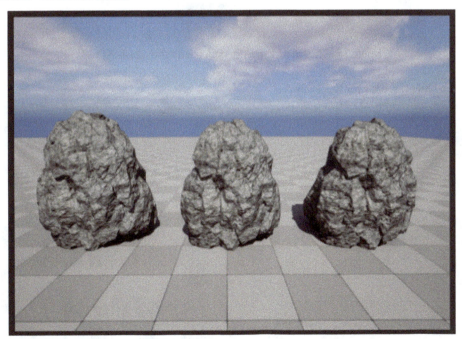

Figure 1.5 – Rock from the Starter Content folder

The following example was created with an actor blueprint, and here is the result for both case uses including **Hierarchical Instanced Static Mesh (HISM)** and **Instanced Static Mesh (ISM)** components that generate 25,000 rocks:

Figure 1.6 – 25,000 rock meshes populated across the level

Here is the blueprint code for this example:

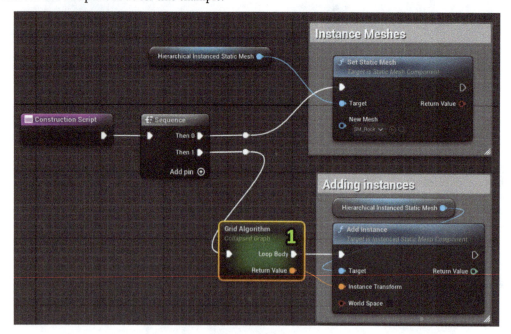

Figure 1.7 – An example of the procedural generation done inside Actor Blueprint

Here is a continuation of the graph that shows the Grid Algorithm, which distributes the static meshes across the level in rows and columns on the X and Y axes:

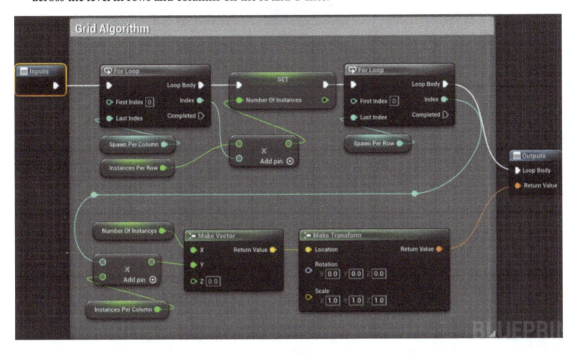

Figure 1.8 – Grid Algorithm for distributing static meshes across the level

Depending on the size of the grid, it may take time to load all the instances. In fact, this whole process takes place on your CPU, and then all the data is sent to your GPU to render your final results on your viewport.

If you want to check out these examples, you can download this project from here:

```
https://github.com/PacktPublishing/Procedural-Content-Generation-
with-Unreal-Engine-5/tree/main/Chapter_1/UE5_PCG_Chapter_01
```

HISM and ISM diagnostics and performance

The following screenshot compares two actor blueprints, one using HISM and the other using ISM. To conduct this experiment, I used Unreal Insights, which is available with Unreal 4 and 5.

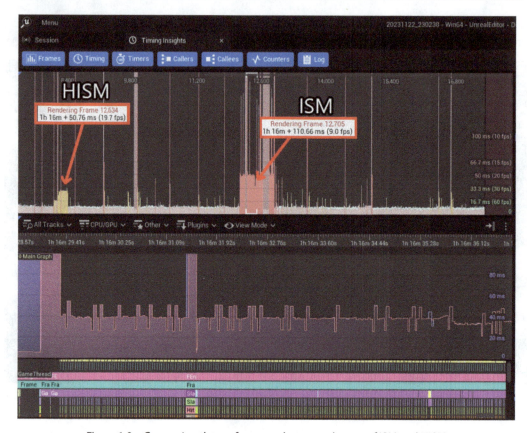

Figure 1.9 – Comparing the performance between the use of ISM and HISM
with Unreal inside Unreal Engine 5

> **Note**
>
> If you're interested in learning how to use the Timing Insights, visit this Unreal documentation website for more information: `https://docs.unrealengine.com/5.3/en-US/timing-insights-in-unreal-engine-5/`.
>
> All of these demos were tested on RTX 3080 with 16 GB RAM and RTX 4090 with 24 GB RAM, and AMD Ryzen 9 7950x with 32 GB and 64 GB RAM.

You can clearly see which one is the winner in terms of rendering and optimization process. Both of those examples are running without Nanite (`https://dev.epicgames.com/documentation/en-us/unreal-engine/nanite-virtualized-geometry-in-unreal-engine`) enabled for the 25,634 rocks mesh asset.

> **Note**
>
> Nanite is Unreal Engine 5's virtualized geometry system, which uses a new internal mesh format and rendering technology to render pixel-scale detail and high object counts

In the next section, we'll explore how the PCG plugin performs by testing it with one of the sample content meshes in the Level

PCG in action

For a more effective comparison between ISM and HISM, let's execute the same example using the PCG tool. As you can see, the rock formation is significantly larger, generating approximately 250,000 rocks. However, the underlying code is notably simpler and highly optimized.

Figure 1.10 – A great example of the power of PCG, which generates 10 times more rocks without a heavy performance load

The following graph is all that's needed to produce the preceding result:

Figure 1.11 – PCG requires few nodes to produce simple results

Clear and straightforward blueprint code has been incorporated into the PCG graph. This pinpoints the nodes that produce the desired results. Remember that PCG is still relying on using HISM or ISM components, but it breaks it down in a more organized and optimized manner.

Here is the performance status for our rock formation, which was running only on PCG Volume:

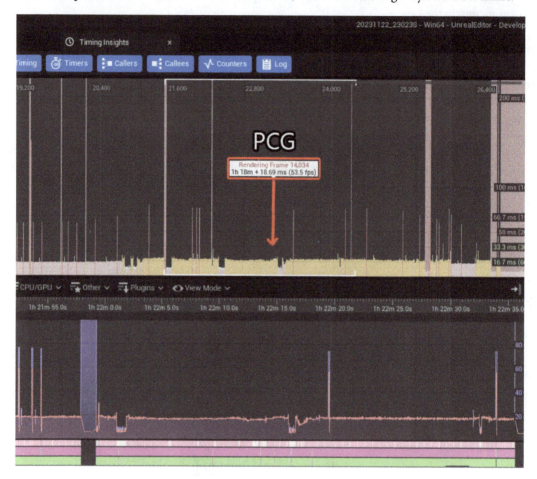

Figure 1.12 – Using PCG doesn't affect performance too much, but it still requires some optimization

As you can see, the results are a lot faster and less heavy for the GPU and CPU usage. This PCG volume is not using Nanite yet. Imagine what a boost of performance this would give you if you have enabled Nanite for this Rock static mesh. We will discuss this in the *Chapter 10* of the book! In this section, we've explored various procedural generation techniques and compared them with the new PCG tool.

In the next section, we will dive into the PCG framework, and we will create our first PCG tool!

Exploring the PCG framework

This book introduces the first steps of creating a PCG graph using a PCG framework. Within this section, I will provide a fundamental explanation of PCG nodes and the overarching structure of PCG graph. I will explain how to recognize the primary components that are essential for constructing the tool.

PCG user interface

To start our journey of setting up the PCG framework, we first need to enable the plugin within the project. I will guide you through the setup process:

1. In the top left corner of your viewport, go to **Edit | Plugins**. In the plugins menu, search for
    `PCG Content Generation Framework (PCG)`. Check the box to enable the plugin, then restart your project.

Figure 1.13 – Enabling the PCG Framework plugin

2. For this exercise, we will use models from the plugin's Content folder, which is located inside the hidden folder of Unreal Engine's project. To access it, we need to access the content drawer by pressing *Ctrl + Spacebar*. Inside your content drawer, go to the top right corner and click the **Settings** button to access the menu.

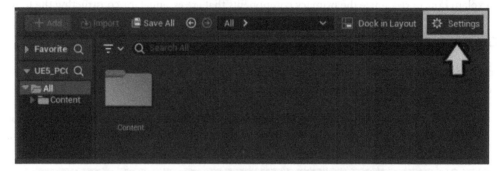

Figure 1.14 – Accessing the hidden folders within the Settings panel

3. Once the menu is open, search and check the **Show Engine Content** and **Show Plugin Content** checkboxes. The Engine folder will appear inside your Content Browser, and you will find the Plugins folder inside the Engine folder.

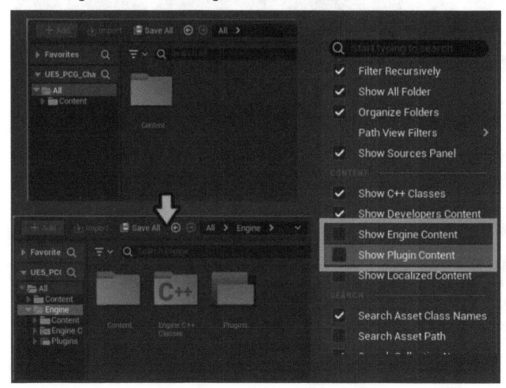

Figure 1.15 – Enabling Show Engine Content and Show Plugin Content

4. Navigate down to the `Plugins` folder and search for `Procedural Content Generation Framework (PCG) Content/SampleContent/SimpleForest/Meshes`. Inside this folder, you will find the assets that we are going to use for our exercise, specifically, the `PCG_Tree_01` static mesh.

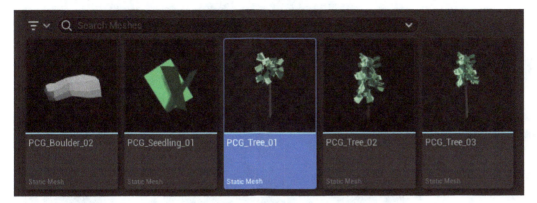

Figure 1.16 – Locating PCG_Tree_01 inside the Plugins folder

In the next section, I will explain the anatomy of the PCG graph user interface and how it functions within Unreal Engine 5.

In this part of the chapter, I will break down all the elements that are needed to create your PCG graph.

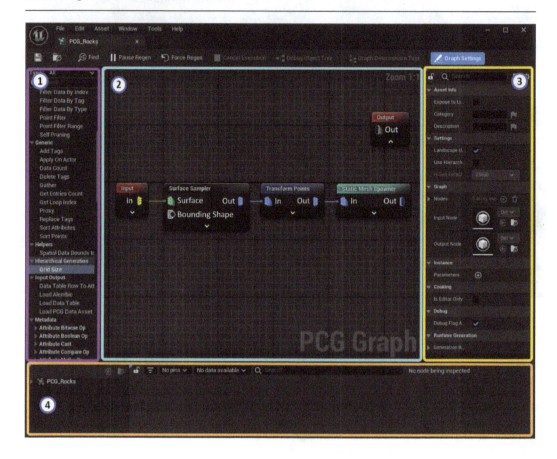

Figure 1.17 – The anatomy of the PCG graph tool

The numbered areas in the preceding screenshot mark different parts of the PCG graph user interface, which I will explain here:

- **1. Node Pallete**: This is where you have the option to select the nodes with which to build the PCG graph. While this is convenient, you can also locate the same nodes by right-clicking on the PCG graph grid space. Nodes represent functions that contribute to constructing a network graph between input and output, which is important for generating outcomes in your environment.

  Each of these nodes carries out a specific operation that contributes to the results, and they are as follows:

  - **Blueprint**: Blueprints, featuring a common node for executing user Blueprints derived from **PCGBlueprintElement**

  - **Control Flow Category**: Nodes that determine the execution order of PCG nodes within a PCG graph, directing how procedural systems generate content

- **Debug**: This section houses nodes that are designed to assist with debugging processes

- **Density**: Nodes found here impact point density

- **Filter**: Nodes in this section filter data based on specified criteria or on a per-point basis

- **Generic**: These nodes are designed to perform basic tasks, such as counting points data, adding or removing tags, and sorting attributes within the PCG graph

- **Helpers**: This node functions as a subgraph that generates points based on the bounds of any spatial data

- **Hierarchical Generation**: This node lets you run a single graph across multiple grid levels, enabling procedural generation at different scales

- **IO**: This section comprises nodes that govern interaction with external data

- **Metadata**: Nodes in this category interact with attributes, whether on points or within attribute sets using mathematical expressions

- **Param**: This section includes nodes that are responsible for controlling the retrieval of parameters from the Actor dataset or attributes

- **Point Ops Category**: This category handles the manipulation of point data by combining, duplicating, splitting, copying, and performing various tasks required to distribute points in different configurations

- **Sampler**: Nodes within this category generate points from a source of spatial data, such as volumes, surfaces, and meshes

- **Spatial**: Nodes here are dedicated to creating spatial relationships between data, altering their internal spatial data, or retrieving data

- **Spawner**: Nodes in this section are responsible for creating new data or placing Actors at specified point locations

- **Subgraphs**: Nodes in this category handle the use of subgraphs, allowing the incorporation of additional PCG graphs into the main PCG graph

- **2. PCG Graph Editor**: This is the primary graph of the PCG editor, enabling you to visually script the PCG structure using nodes. Here, you can drag and drop nodes to create a PCG logic structure that generates environments in your level.

- **3. Details Panel**: This provides you with details about each selected node you are actively working on. You can utilize this information to adjust the values and parameters associated with your chosen node.

- **4. Attributes List**: This section, whether you believe it or not, is important, and it provides in-depth insights into the performance of each spawned mesh. It assesses information for every mesh/actor by evaluating details such as size, scale, color, position, and the density of your spawned static meshes.

Having completed our theoretical discussion of the PCG framework, we will now transition to the practical aspect. In this phase, you will have the opportunity to develop your own PCG tool.

Building your first PCG graph

In the previous sections, we delved into the distinctions between a conventional procedural generation technique employing HISM/ISM and the innovative tool known as PCG. We also explained the fundamental definitions of nodes and their respective functionalities. Now, let's progress further by initiating the construction of our inaugural PCG graph.

Creating a PCG Graph

This time, you will get a chance to construct your own basic PCG graph. Let's start by familiarizing ourselves with the PCG workflow:

1.   Inside your Content Browser, create a new folder and name it PCG.

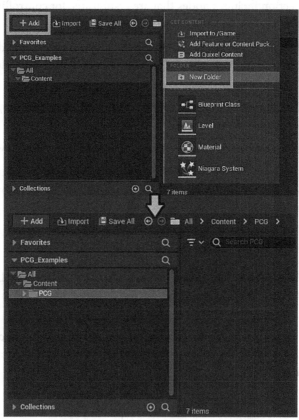

Figure 1.18 – Creating a PCG folder inside the Content Browser

2. Inside the PCG folder, right-click and look for PCG >PCG Graph. Let's rename it PCG_Trees.

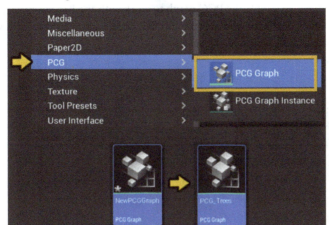

Figure 1.19 – Create a PCG Graph inside the PCG folder and rename it PCG_Trees

3. The next step is to open the PCG Graph by double-clicking on it. You should be able to see the main PCG graph viewport in front of you. This time, we will add a surface sampler by right-clicking on the viewport and searching for Surface Sampler:

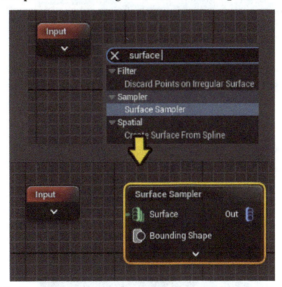

Figure 1.20 – Connecting the Landscape input node with the Surface Sampler
in order for the PCG tool to work

4. With the `Input` node selected, we need to switch **Allowed Types** to **Landscape**. The reason behind this switch is to enable the PCG graph to work with the **Landscape** actor on the scene. On the right-hand side, in the **Settings** panel, click on the arrow next to **Index[0]** and change its **Allowed Types** from **Spatial** to **Landscape**.

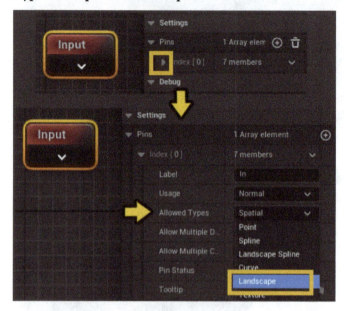

Figure 1.21 – Switching the pin from Spatial to Landscape

At this stage, you can test your tool and see its primary functionality before moving on to the next step. To do this, we need to press the D key to activate debugging mode or go to the details panels with your Surface Sampler node selected.

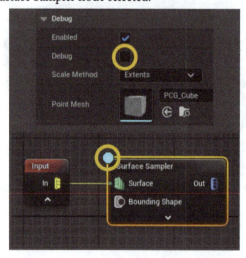

Figure 1.22 – An example of how to activate Debug

5. The blue hexagon above your **Surface Sampler** represents debugging mode being activated on that specific node. Let's go back to our main viewport.

6. Drag and drop your PCG Graph to the viewport scene. You should see a spectrum of cubes from white to black scattered around the landscape surface.

   Debug mode can be seen in the following diagram:

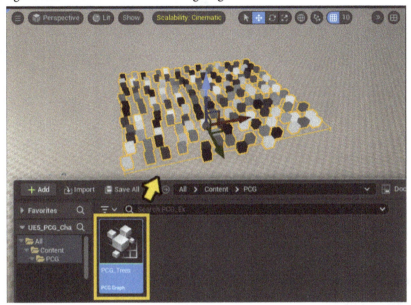

Figure 1.23 – Debug mode enables users to see the points that will be used for spawning all the meshes

   The view shown in the *Figure 1.23* uses grayed out cubes to indicate the position of your static meshes where it will project on the terrain surface

7. Let's go back to the PCG Graph. Inside the PCG Graph, right-click and search for the **Transform Points** node and add it to the graph. Connect the output(**Out**) from the **Surface Sampler** to the **Transform Points** input(**In**) pin.

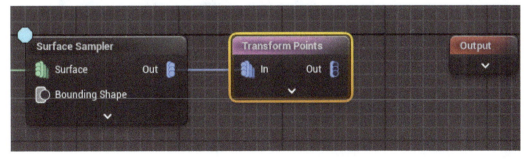

Figure 1.24 – Add the Transform Points node to the PCG Graph

Here are the window settings with the transform information now made visible:

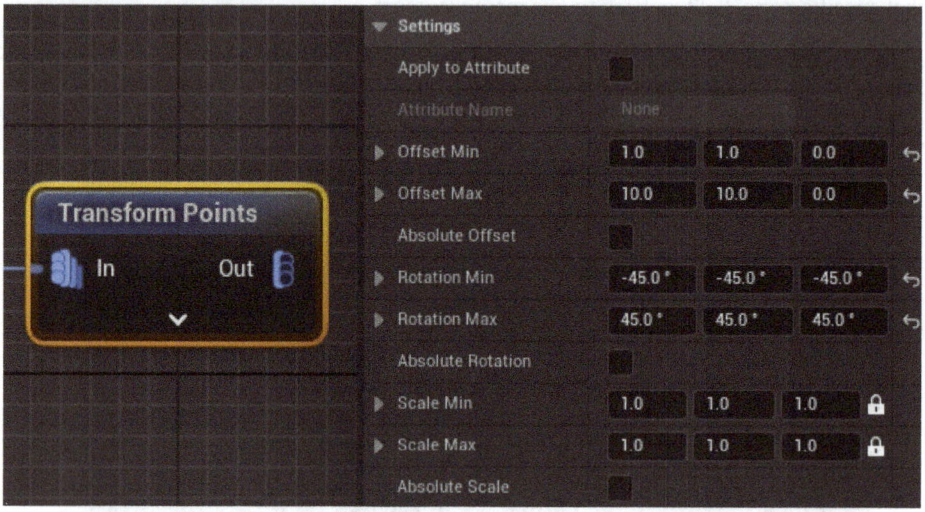

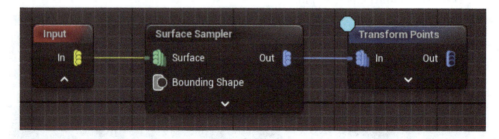

Figure 1.25 – Settings for the Transform Points node in the details panel

8. With the **Transform Points** node selected, you can easily view the details panel on the right-hand side and tweak the values:

- **Offset Min** 1.0, 1.0, 0.0

- **Offset Max** 10.0, 10.0, 0.0

- **Rotation Min** -45.0, -45.0, -45.0

- **Rotation Max** 45.0, 45.0, 45.0

    Let's maintain the current scale of the points for this transform. Make sure to disable debug mode on the **Surface Sampler** node by pressing the *D* key. Then, select the **Transform Points** node and press the *D* key to activate debug mode.

Figure 1.26 – Debugging Transform Points to test the points distribution

> **Note**
>
> Shorcuts are quite useful while working inside a PCG Graph. If you want to review the nodes in debug mode, then you can always press the *D* key on your keyboard. This will allow you to visualize the changes that have taken place on your node.

The results start to take a different shape, and all the cubes are randomly transformed within x and y positions.

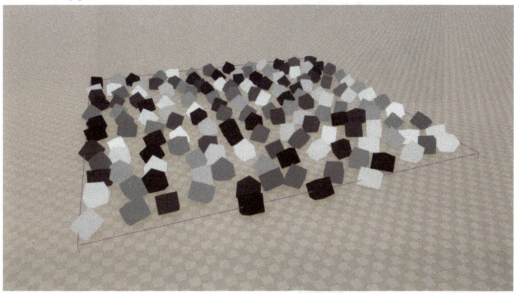

Figure 1.27 – Debugging the random transformation points before spawning the static meshes

9. Last but not least is the **Static Mesh Spawner** node. This is the main part of our graph structure because this is where you spawn the static mesh geometry on the surface of the landscape. Let's turn off debugging mode by pressing the *D* key in the **Transform Points** node, and let's search for our new node by right-clicking on the viewport and searching for Static Mesh Spawner.

Figure 1.28 – Make sure to select the Static Mesh Spawner node

10. With the node selected, navigate to the details panel on the right-hand side. You will find a tab called **Mesh Entities**. Here, we will add a new element to our node.

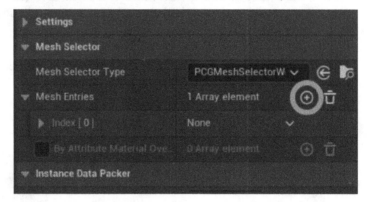

Figure 1.29 – Press the plus sign to add an array element

11. As you can see, your new **Index[0]** has a lot of information about the static mesh that you are about to add to the index array. Of course, you can add more indexes to this node but, for this example, we will only use one mesh entry. Thankfully, the PCG plugin comes with some examples, and here, we can add one of the static mesh geometries to our index list. Let's search for PCG_Tree_01.

Figure 1.30 – Adding PCG_Tree_01 to the Mesh Entry array

12. Let's connect the **Static Mesh Spawner** to the **Transform Points** output pin.

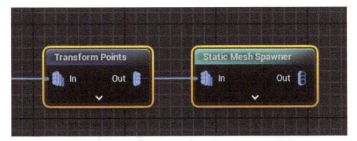

Figure 1.31 – Connecting the Static Mesh Spawner node with the Transform Points node

Once we have set that up, we can view our results in the main viewport.

Figure 1.32 – The outcome of spawning the meshes on the landscape

You have gained valuable knowledge on how to set up your first PCG graph. This knowledge will be very useful when you explore other parts of the PCG graph later on. In the next section, we will go through the other settings that PCG plugin has to offer.

Attributes List

The PCG Graph editor is very handy if you want to analyze a node's attributes and observe the values that are printed into the Attributes Output Log. It usually gives you information about the geometry, the number of meshes being spawned, and their locations.

It's not that difficult to activate this; you just have to press the *A* key or right-click on any node inside your graph and navigate to **Inspect**.

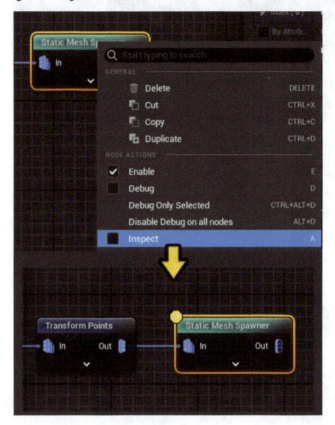

Figure 1.33 – Inspect mode is designed to reveal and display all the details about the meshes that are spawned across the landscape

You will notice the yellow hexagon above the node. This indicates that inspect mode is enabled.

You can view the Attributes output log at the bottom of the viewport and by clicking on the left panel where the PCG component is; there will be a PCG_Trees graph.

Index	Position.X	Position.Y	Position.Z	Rotation.X	Rotation.Y	Rotation.Z	Scale.X	Scale.Y
143	3,605.702	1,756.635	100.048	-0.165	0.323	-0.076	1	1
142	3,330.999	1,694.599	100.04	0.245	-0.306	-0.189	1	1
141	2,900.358	1,760.958	100.035	-0.177	0.17	0.318	1	1
140	2,570.289	1,694.62	100.072	0.209	-0.313	-0.271	1	1
139	2,104.742	1,610.766	100.032	-0.113	0.069	-0.376	1	1
138	1,807.96	1,747.704	100.044	-0.135	0.005	-0.102	1	1
137	1,298.752	1,701.268	100.039	0.072	0.204	0.193	1	1
136	929.155	1,653.303	100.033	0.21	-0.393	0.006	1	1
135	454.834	1,773.715	100.046	-0.175	0.384	-0.08	1	1
134	79.776	1,692.275	100.022	-0.117	-0.076	-0.315	1	1

Figure 1.34 – Attributes window with all the mesh information within the PCG component

We can see that there is a large number of indexes within the PCG Volume space, which is determined by the volume size of the scene. In a nutshell, the bigger the scale of the volume, the more indexes it will hold inside the PCG Volume.

Summary

In this chapter, you learned the value of using the PCG tool and the difference between the existing techniques (that is, HISM/ISM) and the new PCG tool that came with Unreal Engine 5.2.

You've also gained insights into how the performance differs from the existing procedural generation methods and how the new tool can help to process a huge quantity of data and stabilize the overall performance of your levels.

Most importantly, you have learned how to create your first PCG graph by going through the step-by-step methodology of creating a PCG Graph with simple and powerful nodes.

In *Chapter 2*, we will cover more advanced topics. I'll show you how to create a complex PCG graph that will generate a more diverse forest and how it can benefit your workflows.

2

Let's Create Our First Forest!

In *Chapter 1*, we took an overview of buidling a PCG framework from the ground up. We dived into the distinctions between conventional **Hierarchical Instanced Static Meshes (HISMs)** and **Instanced Static Meshes (ISMs)** through blueprints, and we compared them with the innovative PCG tool and its striking difference in performance.

This information allowed us to emphasize the potential of the PCG tool in constructing more complex and diverse environments and structures, laying the foundation for further exploration of its practical applications.

In this practical tutorial, I will show you how to create your own forest environment for gaming or animation purposes. In this project, you will create a procedurally generated forest, filled with realistic trees and vegetation. This approach grants you complete flexibility to modify and tailor various foliage static meshes to suit your preferences.

In this chapter, you will acquire the skills to integrate and merge node structures within a single PCG graph. The tutorial gives you an opportunity to combine multiple trees while incorporating variations of grass, bushes, and stones across a landscape. In this exercise, we will use Megascans assets, available for free on Quixel and the Unreal Engine Marketplace.

Subsequently, you will learn how to combine all these assets as one unified PCG graph, offering you control over density adjustments based on your preferences.

The main topics covered in this chapter include the following:

- Gathering and optimizing Megascans models
- Building a PCG graph structure
- Combining and merging nodes in a PCG graph
- Finalizing a unified PCG graph

Upon finishing this hands-on tutorial, you'll have a deeper understanding of the PCG workflow and its tool setup. This knowledge will help you to create your own PCG tools to design environments and levels within Unreal Engine. Now that you're familiar with the content we'll explore in this practical tutorial, let's delve into the Unreal project that I've prepared for you to download.

Technical requirements

You will need the following hardware and software to complete this chapter:

- A computer that can run a basic project with at least Unreal Engine 5.

- The latest Quixel plugin installed in Unreal Engine. For more information, go to `https://docs.unrealengine.com/5.0/en-US/quixel-bridge-plugin-for-unreal-engine/`.

The Unreal Engine version that's used in this chapter is version UE 5.4. The PCG tool was introduced with Unreal Engine 5.2, but there are some nodes that got deprecated in Unreal Engine 5.4; hence, the latest version of Unreal Engine is the most beneficial for this chapter.

You can download this project from the following GitHub link:

`https://github.com/PacktPublishing/Procedural-Content-Generation-with-Unreal-Engine-5`

The code in action video for the chapter can be found at `https://packt.link/0KjQk`

Downloading assets from Quixel

Before we can begin our journey to the practical tutorial, I would like to explore the working files for this tutorial, which we need to add from the **Marketplace**. All the assets are free, and you just need to use your Epic Games account to log into your Epic Launcher.

Quixel Bridge is a plugin for Unreal Engine that gives you full-featured access to the Megascans library within the Level Editor. You can browse collections, search for specific assets, and add assets to your Unreal Engine projects.

1. Open your Epic Launcher, navigate to the **Marketplace** tab, and then search for `Megascans Trees: European Beech` (see *Figure 2.1*). Add them to your project.

2. **Megascans Trees: European Beech** is the asset you need in the Marketplace, and its size should be around 6.9 GB.

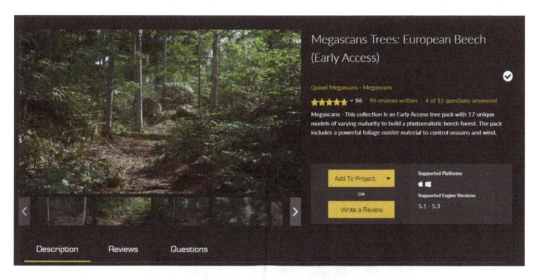

Figure 2.1 – Adding an asset from the Marketplace

> **Note**
>
> The Quixel Megascan trees are currently not compatible with Unreal Engine 5.4, but they will be available soon. In the meantime, I've put together a tutorial on how to migrate the Megascan Trees: European Beech assets into Unreal Engine 5.4. For further details, check the link below: https://www.youtube.com/channel/UC-WefInHnQJkcxfEYQJHCew

3. The next step is to download other assets. We will use some rocks from Quixel Bridge, which we can import into the project with a **Bridge plugin**. At the top of your viewport, click on the + button and then **Quixel Bridge**.

Figure 2.2 – The Quixel Bridge option

4. On the search bar inside Quixel Bridge, search for the `rock` keyword, and then pick a rock asset that you would like to work with. For this exercise, I chose **NORDIC FOREST CLUSTER ROCK LARGE**.

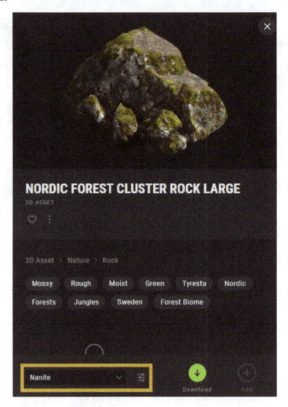

Figure 2.3 – Setting up the import settings to Nanite

Make sure to change your **Import** settings to **Nanite** from the drop-down box, as this will save you lots of time later when we start optimizing our assets!

After you have downloaded all the assets to your project, we will move on to the next part of the tutorial, which is preparing and optimizing our mesh data.

Optimizing the data for PCG use

With all the assets prepared in your project, it is important to prepare your mesh data and make it bulletproof, saving your project from any performance bottlenecks. In this case, we will use the following checklist to make sure that we won't be experiencing any issues while spawning the static meshes across a landscape.

Figure 2.4 – The tree model location inside the PivotPainter folder

1. Inside the content browser, navigate to **EuropeanBeech | Geometry | PivotPainter**.

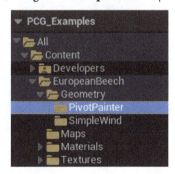

Figure 2.5 – Navigating to the PivotPainter option

Figure 2.6 – The tree models to use for this example

2. In this tutorial, you can choose any tree mesh you want, but in my case, I chose trees from the selection shown in the preceding screenshot.

3. With the three tree models selected together from the `PivotPainter` folder, *right-click* on them and choose **Nanite**.

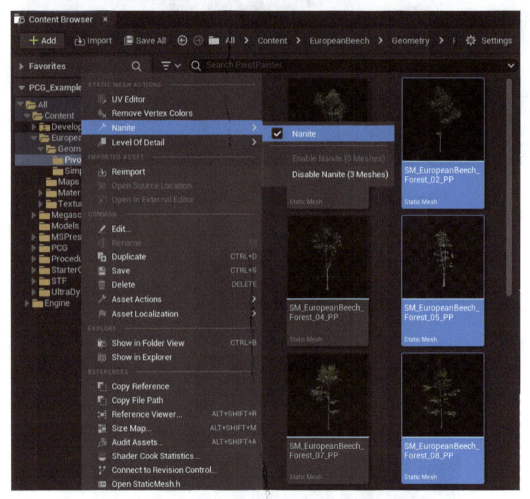

Figure 2.7 – Turning on Nanite for the trees

4. The next step is to enhance the settings by preserving the area of the foliage, ensuring that it will be much sharper and visible in the distance. *Double-click* on any of the trees that you will be working with, and then go to the **Details** panel. Inside the mesh editor on the right-hand side, navigate down to **Nanite Settings** and tick the **Preserve Area** box.

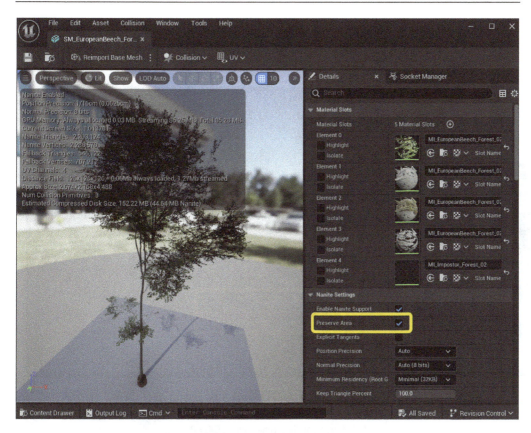

Figure 2.8 – Opening and inspecting SM_EuropeanBeech_Forest_02

Activating these settings within the model ensures that it maintains stable performance throughout its use in the exercise.

> **Note**
>
> The Quixel Megascan trees can be too heavy to run on lower-end computers, even with the **Nanite** enabled! I would suggest using alternative tree models, which you can download from the Marketplace. I will explain this topic further in the *More free assets from the Marketplace* section later in the chapter.

5. Make sure to repeat the same process for every tree.

6.   Last but not least is the tiniest asset in the tree Megascans package, SM_EuropeanBeech_
     Seedling_04_PP, which we will use together with the grass spawner in the Getting started
     with the PCG graph section later in this tutorial.

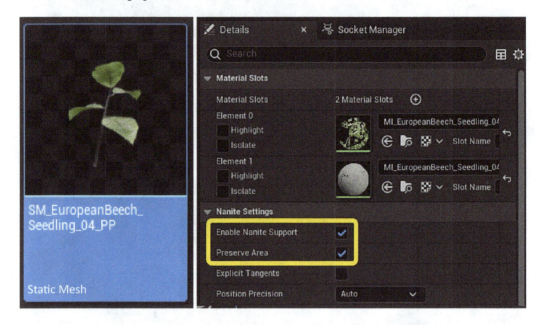

Figure 2.9 – Ticking the Enable Nanite Support and Preserve Area boxes for
SM_EuropeanBeech Seedling_04_PP

7.   Make sure to tick **Enable Nanite Support** and **Preserve Area** inside the mesh settings.

As we delve into the enchanting world of (**PCG**) **Procedural Content Generation** within Unreal
Engine, it's crucial to acknowledge the treasure trove of resources available at our fingertips.

More free assets from the Marketplace

Surprisingly, there are lots of very awesome assets that you can get for free from the Unreal Marketplace.
One of them, which we are going to use, is **Landscape Pro 2.0 Auto-Generated Material**. It consists
of a cool grass foliage mesh, which we can use for our PCG forest environment.

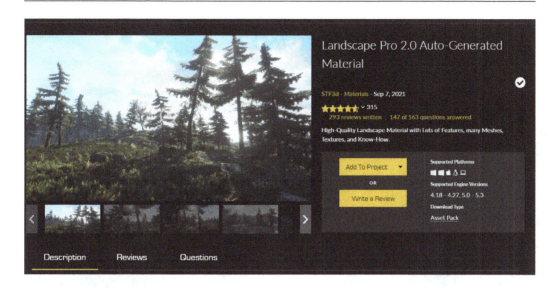

Figure 2.10 – Adding the Landscape Pro 2.0 Auto-Generated Material package to the project

Having navigated the vast array of resources in the Unreal Engine Marketplace and selected **Landscape Pro 2.0 Auto-Generated Material** for our project, the next step is to integrate this powerful library into our environment:

1. Once you have added **Landscape Pro 2.0 Auto-Generated Material** to your project, inside the content browser under **STF | Pack03-LandscapePro- | Environment | Foliage | GrassBush**, search for SM_grass_bush04_lod00.

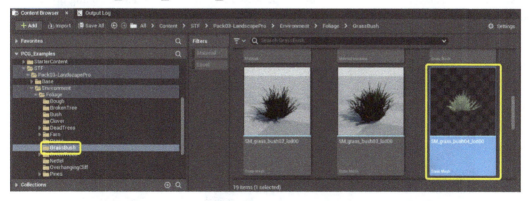

Figure 2.11 – Using the SM_grass_bush04_lod00 asset for this exercise

2. *Double-click* on that asset and repeat the process of ticking **Enable Nanite Support** and **Preserve Area**, as we did with the previous tree models.

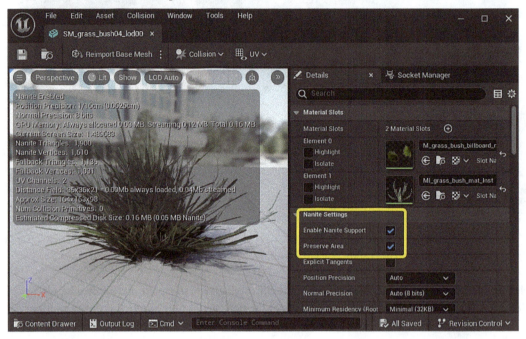

Figure 2.12 – Enabling Nanite Support and Preserve Area for SM_grass_bush04_lod00

Unsurprisingly, Quixel trees can be heavy on PC performance, and I recommended being prepared with some alternative assets before we move on to the practical exercise together. We will use the tree assets from the same free materials that we downloaded to use the **Landscape Pro 2.0 Auto-Generated Material** grass assets.

Figure 2.13 – Selecting trees from Landscape Pro 2.0 Auto-Generated Material

In the next section, we are going to set up the landscape that we need for our PCG setup to work.

Setting up a Landscape

It is important to add a very simple landscape, which is needed for the PCG volume to function within a scene:

1. In the top-left corner of the viewport, click on **Selection Mode** and search for Landscape. Then, select the **Landscape** mode.

Figure 2.14 – Selecting the Landscape tool

2. In the **Manage** tab of your Landscape tool, scroll down to the bottom and click the **Create** button. This will generate a new default Landscape actor.

Figure 2.15 – Creating a landscape actor

3. Select your new landscape actor in the outliner. In the Details Panel, search for Landscape Material and add the **M_Ground_Moss** material to the **Landscape Material** slot.

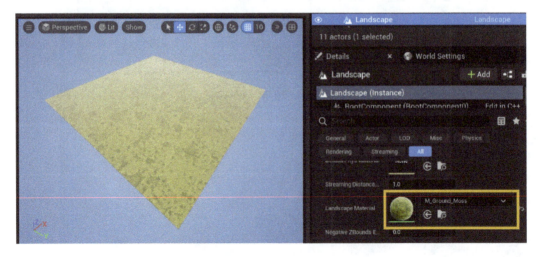

Figure 2.16 – Assigning the M_Ground_Moss material to our Landscape actor

We have completed setting up the landscape for our project. In the next section, we will start constructing our first PCG graph, which will help us create the desired forest.

Getting started with the PCG graph

We have optimized all the assets needed for our practical tutorial, and now, we are ready to create a PCG tool.

For this particular exercise, we will use a GitHub repository (`https://github.com/PacktPublishing/Procedural-Content-Generation-with-Unreal-Engine-5`) where you will find a level with the landscape already laid out for this tutorial!

1. Now, let's look at the steps for this tutorial, as follows:

2. First, *right-click* on your folder space inside the `Content` folder, and search for **PCG | PCG Graph**. Press *F2* on your keyboard and rename the PC Graph `PCG_Forest`. As you can see, I have organized my folder structure based on the chapters of this book. Feel free to do the same at your end.

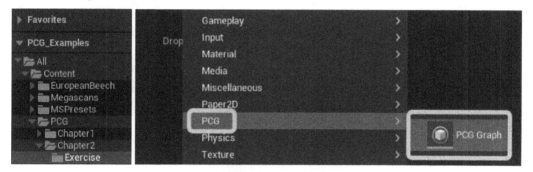

Figure 2.17 – Creating a PCG graph under Content | PCG | Chapter2 | the Exercise folder

3. *Double-click* on your newly created PCG graph and start adding a few elements to the graph. For now, we'll create comments for each network graph, using them as a placeholder to indicate which sub-graph is responsible for each mesh that is distributed to your landscape.

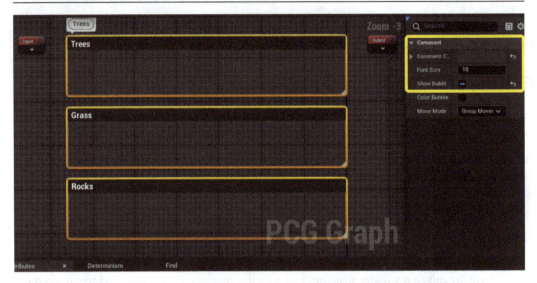

Figure 2.18 – Marking the areas designated to construct the PCG graph network for each foliage spawner

4. Press the *C* key button on your graph to create new comments, and rename them Trees, Rocks, and Grass, respectively. You can change the color of your comment to whatever you prefer by going to the **Details** panel on the right-hand side.

5. *Right-click* in space, and search for Surface Sampler. Duplicate it three times with the *Ctrl + D* keys, or *right-click* on the node and select a **Duplicate** command.

Figure 2.19 – Adding a surface sampler for each designated area

Rename your Surface Sampler to something else, which is quite easy to do! First, make sure your node is selected.

Figure 2.20 – Renaming the Surface Sampler

6. Click on the node once again and then the selected node will enter *rename mode*. I named this node SS Trees, which stands for **Surface Sample Trees**.

Note

If you want to, you can repeat the same process for the other nodes, but this is not really necessary and is up to you.

---

7. Let's add Transform Points to our graph and place them next to each Surface Sampler. The **Transform Points** node will be used to customize the transformation position of each procedurally generated static mesh within a scene.

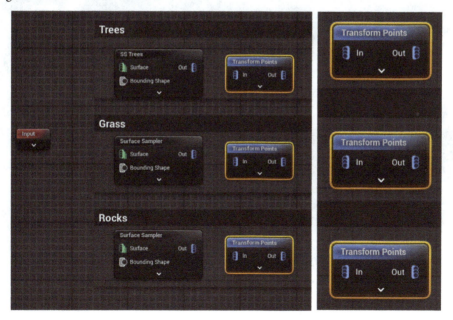

Figure 2.21 – Adding a Surface Sampler to each designated area

8. For now, let's focus on the **Trees** graph network, to which we need to add a few more new nodes. But before we do that, let's add **Static Mesh Spawner** and connect it to the **Transform Points** node. Using the **Static Mesh Spawner** node, you can spawn your static meshes at the positions defined by the points from the **Transform Points** node.

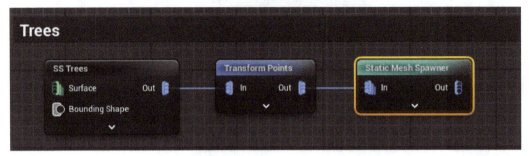

Figure 2.22 – Connecting Static Mesh Spawner to the PCG graph

9. Before connecting the Input node directly to the Surface sampler, an input pin must be added to enable mesh generation on the landscape surface. To do this, navigate to the Settings panel located on the right side after selecting the Input node. Add a new index array, set its Allowed Type to Surface, and rename it to Landscape. This will display the surface landscape as a green input pin on the Input node.

Figure 2.23 – Adding a Surface input inside the Input node and renaming to Lanscape

10. Now, let's test our first iteration. Connect the **Input** node to the **SS Tree** node, and let's debug our **Static Mesh Spawner** for the **Trees** sub-graph by pressing the *D* key button. Debug mode provides visual and, sometimes, quantitative feedback on the operations and outputs of the PCG graph.

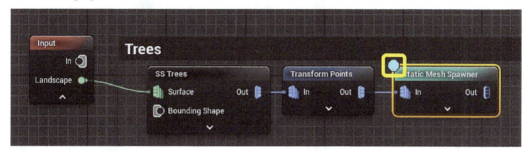

Figure 2.24 – Enabling debug mode over Static Mesh Spawner

11. Let's *drag and drop* a PCG_Forest graph into the scene and test out the results. Your graph will generate points that are rendered under **Debug** mode; hence, you will see black and white cubes scattered across the scene.

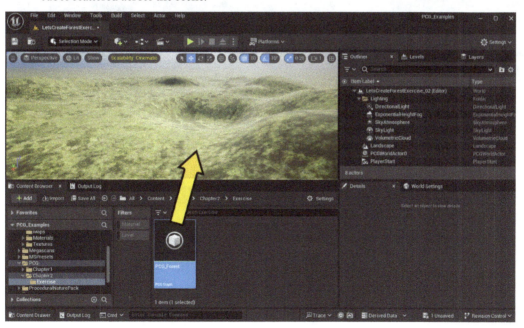

Figure 2.25 – Dragging and dropping a PCG graph to test it on the scene

12. As you can see, the boxes appear in shades of gray, black, and white! This exemplifies the activation of **Debug** mode, showcasing the arrangement and distribution of PCG points throughout a landscape.

Figure 2.26 – Debugging Static Mesh Spawner in the scene

13. For a better view, let's rescale the selected PCG volume slightly so that we can have a better view of the PCG distribution across the landscape.

14. Here is the scale that I chose for the PCG volume. You can make it slightly bigger for your preferences.

Figure 2.27 – Scaling the PCG volume on the details panel

> **Note**
>
> This is the same project that you can download from the GitHub link that I provided at the beginning of this chapter. It consists of a very simple landscape setup and the landscape materials from the `StarterContent` folder. This is enough for our exercise, and you can modify or redesign the landscape in any way you feel comfortable.

In the following section, we will go deeper into enhancing the **Static Mesh Spawner** node by incorporating our prepared models.

Modifying the Static Mesh Spawner

Up to this point, we've incorporated the crucial nodes into our graph, allowing us to debug and observe the distribution across the landscape through the nodes network. Now, let's shift our focus to integrating the static meshes of our trees into **Static Mesh Spawner**.

Let's revisit the PCG graph (specifically, **Static Mesh Spawner**), and then add some important elements that will prepare us for the next creative steps:

1. Let's go back to the PCG graph and turn off **Debug** mode on **Static Mesh Spawner node**. Select **Static Mesh Spawner** again, go to the right-hand side under **Mesh Entries**, and start adding three array elements by clicking the + button in the right-top corner, located to the right of **Mesh Entries**.

Figure 2.28 – Adding mesh entries to each array inside Static Mesh Spawner

2. This will allow us to randomize three different trees and add a different variety to our landscape.

3.   Let's open each of the entries and start adding the trees that we chose and optimized for this exercise:

•   First, we'll look at how we can add Megascans tree models to the **Mesh Entries** array for each Megascans Trees entry:

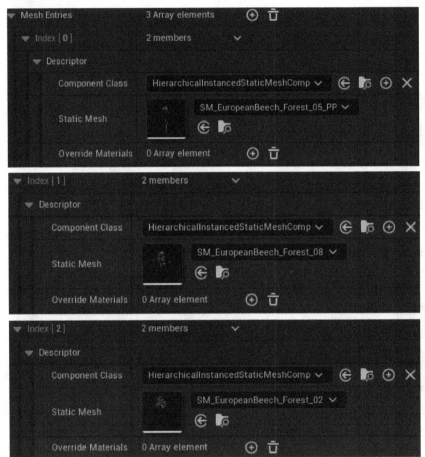

Figure 2.29 – Adding trees from the Megascans

•   Incorporating Megascan trees into your project can significantly impact your PC's performance by consuming your CPU and RAM Memory. As an alternative, consider using the optimized trees provided by the **Landscape Pro 2.0 Auto-Generated Material** package for better efficiency.

- Now, we'll look at how we can add **alternative trees**:

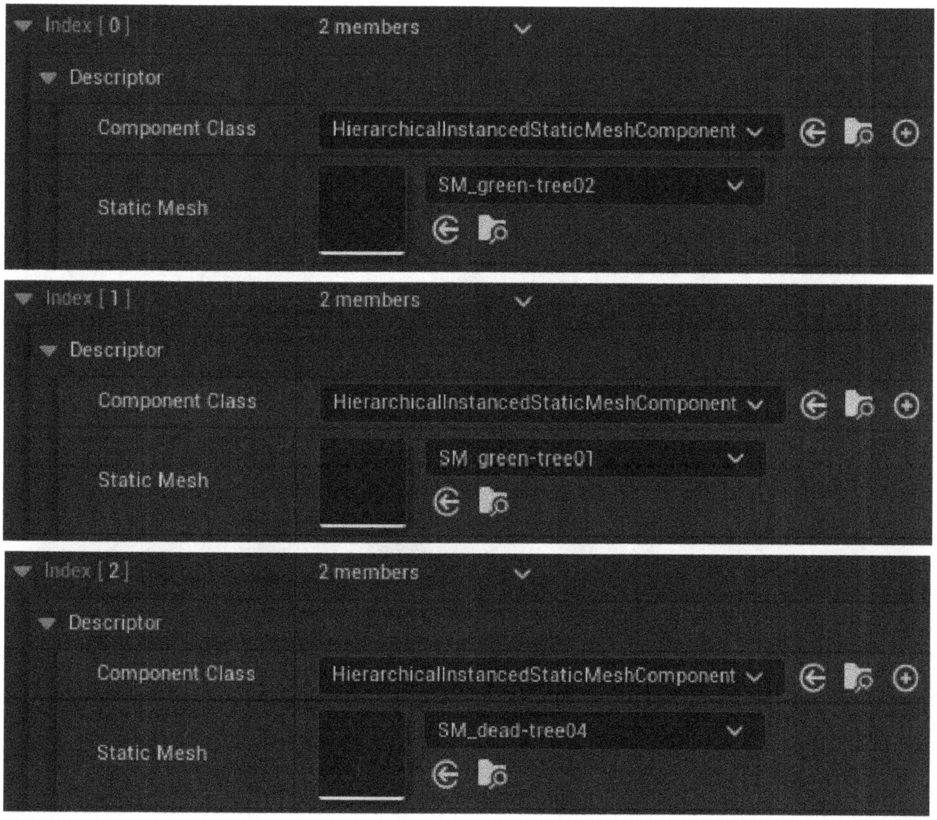

Figure 2.30 – Adding trees from the Landscape Pro 2.0 Auto-Generated Material package

4. The trees from the **Landscape Pro 2.0 Auto-Generated Material** package enable you to adjust and expand the number of assets that can be spawned in the scene. You may have observed a noticeable improvement in performance compared to when using Megascan Tree assets.

Note

As you can see, by choosing each **Index**[] array for a tree in the **Descriptor** tab, you gain the ability to customize and optimize the static mesh, enabling further adjustments. This option is particularly useful if you intend to enhance project performance, possibly at the expense of reducing shadow or lighting quality.

5.   Go back to your main viewport, and you should see the results on your screen!

Figure 2.31 – Trees spawned within the PCG volume

Something is not right, and it looks way too busy for this particular example. Resizing a PCG volume will not help here, so in this case, we have to go back to our graph and modify it slightly.

6.   Inside the PCG graph, select the Surface Sampler for the trees (**SS Tree**), and increase the **Point Extents** number for each axis at **X** and **Y**, leaving **Z** at **100**. Increase the **X** and **Y** values from **100.0** to 200.0.

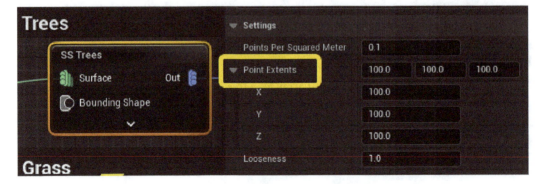

Figure 2.32 – Increasing the Point Extents values for the Surface Sampler

7.   With this part completed, you should see your trees distributed evenly with less space density!

Figure 2.33 – Enhancing the Point Extents values leads to a wider spacing of the trees' positions

> **Note**
>
> Even though PCG is an ideal solution for level design, it still requires constant supervision, especially when considering performance! Make sure to increase the **Point Extents** values if you want to achieve great results with an optimal frame rate.

The trees appear a bit strange because their roots are projected based on the shape of the landscape.

Figure 2.34 – Trees being spawned at the wrong angle

In this case, we must change the rotation settings and set them to **Absolute Rotation**. Let's go back to our PCG graph and select the **Transform Points** node. With this selected, navigate to the right-hand side and tick the **Absolute Rotation** box.

Figure 2.35 – Enabling Absolute Rotation inside Transform Points

As you can see, this has fixed our issue, and now, we can modify the trees' transformation range and their offset position.

Figure 2.36 – Fixing the trees' angle rotation

Let's look at the definitions and functions of each property that accompanies **Transform Points**:

- **Offset**: Modifying this configuration will allow the displacement of each tree to occur within random values, specified by the **Min** and **Max** range. Higher **X** and **Y** values, depending on your landscape, will determine the spacing between trees – closer together or farther apart. The **Z** value will dictate whether each tree is positioned lower or higher, relative to the ground floor.

- **Scale**: With this setting, you can randomly change the **X**, **Y**, and **Z** scale values of your tree accordingly across a landscape. This is particularly useful when you have different types of plants and vegetation.

- **Rotation**: This is a rotational configuration for a spawned mesh. This setting comes with the **Rotation Min** and **Rotation Max** settings. Don't go too crazy and limit the adjustments, perhaps just setting values within the range of -10 to 10 degrees for both the **X** and **Y** axes. For the **Z** axis, consider a broader range, ideally setting **Min** to -180 degrees and **Max** to 180 degrees. This configuration will introduce random rotations along the **Y** axis for the trees.

We have completed the setup of the **Transform Points** node. In the upcoming section, we will begin incorporating various foliage static meshes into the forest environment.

Adding various vegetation to the forest

In the previous part of our exercise, we implemented the tree network, and we now know how to manage nodes accordingly so that we can create our first forest. However, a forest doesn't only consist of trees, right? Let's shift our focus to adding other assets, such as grass and rocks, to our forest. Let's get started by adding some grass.

Adding grass to the forest

We optimized some grass foliage at the beginning of this chapter in the *Downloading assets from Quixel* section, and now, we can use it and implement it in our PCG graph network as a new asset for the **Static Mesh Spawner** node.

Let's start adding grass meshes to the mesh entries:

1. Click the *right mouse button* on the PCG graph viewport. Search for `Static Mesh Spawner`, and then add it to the graph.

2. With **Static Mesh Spawner** selected, go to the **Details** panel on the right-hand side and add two mesh entries.

Figure 2.37 – Adding two Mesh Entries to the Static Mesh Spawner node

3. Search for SM_grass_bush04_lod00 and SM_EuropeanBeech_Seedling_04_PP, which we used to prepare the previous exercise, as follows:

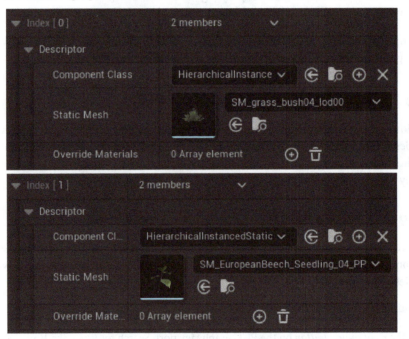

Figure 2.38 – Adding SM_grass_bush04_lod00 and SM_EuropeanBeech_Seedling_04_PP to each array

4.   Connect the **Input** node to each existing node, as follows:

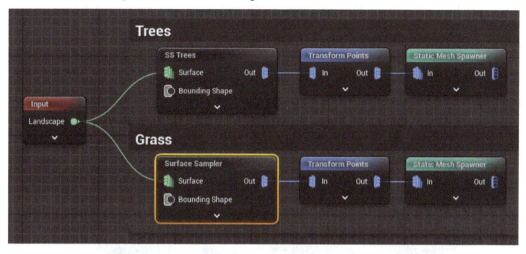

Figure 2.39– Connecting Input to each existing node inside the PCG graph

Now, let's go back to our main viewport and check out the results!

Figure 2.40 – Testing the results of the spawned grass distribution

Currently, there is only a little grass spread across the landscape, which is expected, since we still need to tweak the distribution under **Points Extents** inside the **Surface Sampler** node. Let's head back to our PCG graph and change the **Points Extents** and **Transformation** settings inside the **Surface Sampler**, **Transform Points**, and **Static Mesh Spawner** nodes.

Surface sampler settings

To attain a lush landscape with abundant grass, it is necessary to decrease the spacing between each grass point spawned on the terrain. The most straightforward approach is to decrease the values for **X**, **Y**, and **Z**, and change the density of points per square meter. Let's go back to our PCG graph and adjust those settings.

Points Extents

Inside the PCG Graph viewport, select the **Surface Sampler** node and navigate to **Settings** on the right-hand side. Change the **X**, **Y**, and **Z** values from **100.0** to 20.0.

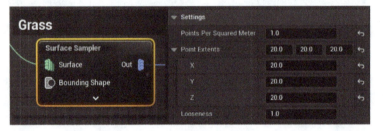

Figure 2.41 – Decreasing the Point Extents values to increase grass distribution across the landscape

Make sure the **Points Per Squared Meter** value is is set between 0.1 and 1.0!

Figure 2.42 – The grass density has increased by a factor of 10

The grass is now spawning in positions that are more widely spread out from each other.

Transform points settings

Things are improving, but we have not finished yet, and we still have to add a few more settings inside the different nodes. As you can see in the following screenshot, in some locations, the grass is spawned in lines parallel to each other, which usually means that points are missing a random transformation.

Figure 2.43 – Examining the issue with the grass rotation

To start our configuration, we have to work on the displacement of the grass spawned at each point on the landscape terrain:

1. First, let's go to the **Transform Points** node, where we are going to change the transformation values for a better and more realistic grass distribution.2. Select **Transform Points** and check the **Settings** on the right-hand side. There, we can add the values that we need for the points transformation.

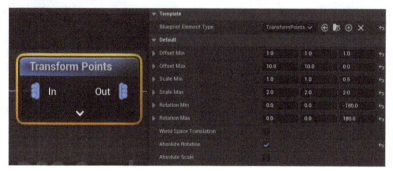

Figure 2.44– Setting the rotation values for the Transform Points node

2. For better legibility of the values in the preceding screenshot, I have created the following table with the values that I used for this exercise:

Transform Points	X	Y	Z
Offset Min	1.0	1.0	1.0
Offset Max	10.0	10.0	0.0
Scale Min	1.0	1.0	0.5
Scale Max	2.0	2.0	2.0
Rotation Min	0.0	0.0	-180.0
Rotation Max	0.0	0.0	180.0

Table 2.1 – Transform Points and the corresponding values for the X, Y, and Z axes

3. The last setting involves ticking **Absolute Rotation** below our transformation settings (see *Figure 2.44*).

Now, let's check out the results! You can see how the grass is starting to distribute evenly:

Figure 2.45 – The grass rotation fixed

In this section, we worked out the spawning distribution using the transform points, and we configured values to improve the distribution quality over the terrain. Now, we will take a look at our last node, **Static Mesh Spawner**, and we will change the **Weight** settings under the mesh entries for the grass static meshes!

The Static Mesh Spawner settings

The **Weights** setting is a setting that calculates how many meshes will be spawned equally, depending on the points distribution across a landscape. In this case, we have two types of grass that are spawned across our terrain simultaneously, but one grass type has more density than the other, which makes them unequally distributed across the landscape. We can fix this by simply changing the weight of both grass static meshes.

Let's look at the steps to change the weight of the two grass static meshes:

1. With the **Static Mesh Spawner** node selected, navigate to the right-hand side and look for the **Mesh Entries** panel; here, **Weight** is set to **1** by default.

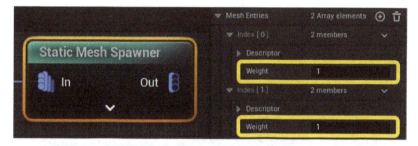

Figure 2.46 – The Weight location for each mesh entry

2. Let's change the numbers randomly for each of the mesh entries, and in this case, we want to populate one type of grass more than the other. To do this, we will simply enter the values of 200 and 100 in each **Weight** field.

Figure 2.47 – Increasing the weight for each mesh entry to equally distribute the foliage in the landscape

3. Currently, as shown in the following screenshot, we can see one type of grass, but we can't really see much of the other type. To fix this, we can tweak the **Weight** settings again and balance the grass types.

Figure 2.48 – The results after increasing the number of weights

4. Let's see what happens if we swap the numbers.

Figure 2.49 – Swapping the number of weights to test different results

Now, we get a bit more random foliage across the terrain, which helps to make the forest ground very diverse rather than just one type of grass.

Figure 2.50– Grass distribution results

So far, you've acquired the skills to work with the appropriate settings for each node, applying that knowledge to create a basic functional solution for the procedurally generated forest. The final step involves distributing rocks across the landscape using the techniques covered earlier.

Adding rocks to the forest

Before we can move on to the most exciting part of the exercise, it would be great to add some rocks to our terrain and make them appear like they are part of the forest landscape. To do this, let's dive into our PCG graph once more and add extra nodes:

1. Let's repeat the same process as before by *right-clicking* on the PCG graph viewport space and searching for `Static Mesh Spawner`.

2. Connect this node to the **Transform Points** node so that we have it ready for testing. Add only one entry under **Mesh Entries**.

Figure 2.51– Implementing mesh entries for Static Mesh Spawner

3. For this example, we will use the rock shown in the following screenshot. Expand the **Mesh Entries** array and search for the `S_Mossy_Boulder_xetmcci_high_Var1` rock asset.

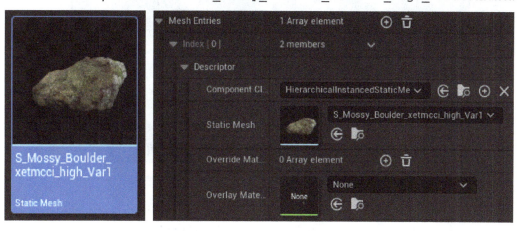

Figure 2.52 – Adding S_Mossy_Boulder_xetmcci_high_Var1 to the mesh entry array

4. Before we can move on to the final stage, we have to change the **Point Extents** values on the **Surface Sampler** node. The following screenshot shows the numbers that I think will work

well for the rock distribution. The higher the numbers, the more the rocks will be dispersed to create a natural displacement.

Figure 2.53 – Modifying Point Extents for the Rocks distribution

5. The last step is to set the **Transform Points** settings and enable **Absolute Rotation**.

Figure 2.54 – Adding and modifying values to Transform Points for a better rock formation

6. For better legibility, I have created the following table with the values that I used for this exercise. You can copy these values and paste them into your **Transform Points** node:

Transform Points	X	Y	Z
Offset Min	0.0	0.0	0.0
Offset Max	0.0	0.0	0.0

Transform Points	X	Y	Z
Scale Min	1.0	1.0	1.0
Scale Max	2.0	2.0	2.0
Rotation Min	0.0	0.0	-180.0
Rotation Max	0.0	0.0	180.0

Table 2.2 – Transform Points and the corresponding values for the X, Y, and Z axes

Now, let's test our PCG graph and see the outcome of it in the main viewport.

Figure 2.55 – Testing the results of the rock formation

As you can see, we are starting to get some interesting results with the rock distribution, and it would be interesting to see whether we can add another extra feature to end this tutorial. How about adding some grass/moss on top of the rocks? For this example, let's reuse some existing grass from the previous sections.

Improving a forest PCG tool

As you saw in the previous screenshot, the current configuration allows the trees to spread evenly, but that causes the rocks to overlap with each other. The solution involves adding a configuration that reduces the variance in the placement of spawned trees across the landscape, filling the gaps with static mesh rocks. This solution will be a good scenario in which to use the PCG tool. It will require a few extra nodes, as follows:

- **Self Pruning**: This node proves valuable in eliminating any overlapping points that may obstruct the placement of other static meshes interacting with the terrain during spawning.

Figure 2.56 – The Self Pruning node

- **Difference**: This node helps to eliminate points from another set that represents trees and rearrange them in an organised order. In other words, it differentiates the positions of the points, depending on the static mesh spawning a position on the terrain.

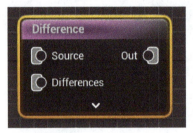

Figure 2.57 – The Difference node

- **Bounds Modifier**: This node comes in very handy. It applies a transformation on the point distribution and its boundary. This can be utilized to establish boundary limits and adjust the area range of points.

Figure 2.58 – The Bounds Modifier node

In the next phase of the exercise, we will configure extra nodes within the PCG graph to diversify the forest.

Setting up nodes inside the PCG Graph

Let's start connecting these new nodes in a correct orderr:

1. In the PCG graph, navigate down to the **Rocks** network graph, *right-click*, and search for Self Pruning. Connect a **Transform Point** node directly to the **Self Pruning** node. There is no need to tweak anything in this node, as it comes complete with the packed settings that we will use for this configuration.

Figure 2.59– Adding the Self Pruning node to the Rocks PCG graph

Right-click on the viewport space again, and this time, search for Difference. Connect the **Self Pruning** node to the **Difference** node, as shown in the following screenshot. Make sure you are connecting to the **Source** output!

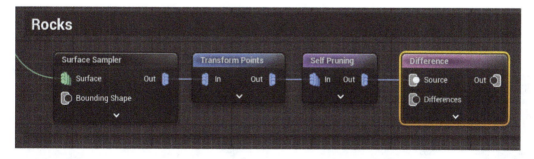

Figure 2.60 – Adding the Difference node to the Rocks PCG graph

Now, we have to make some minor tweaks to this node. With the **Difference** node selected, go to the **Details** panel on the right-hand side, and under **Density Function**, choose **Binary**.

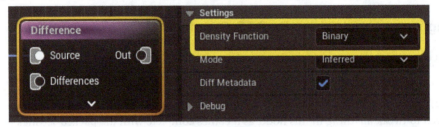

Figure 2.61 – Modifying the Difference node and setting Density as Binary

2. We'll leave the **Rocks** graph network for now and come back to it later! Let's move on to the **Trees** graph network, *right-click* on the graph, and search for Bounds Modifier.

Figure 2.62 – Modifying values inside the Bounds Modifier node

> **Note**
>
> You can opt for any other values, but let's use these values for now in this exercise!

3. Connect the **Transform Points** node to the *input* **Bounds Modifier** node.

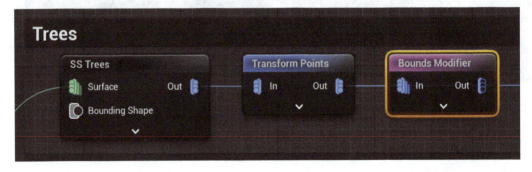

Figure 2.63 – Connecting the Bounds Modifier node to the Transform Points node

4. Now, we need to connect the *output* of the **Bounds Modifier** node to the **Differences** *input* of the **Difference** node, as shown in the following screenshot:

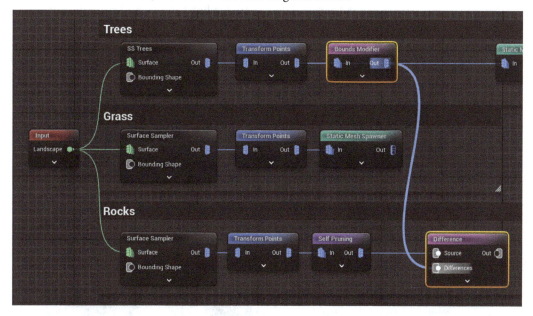

Figure 2.64 – Connecting the Bounds Modifier node to the Difference node

5. Make sure to connect the Static Mesh Spawner to the Transform Points node., Keep an eye on the connection between the **Transform Points**, **Bounds Modifier**, and **Static Mesh Spawner** nodes. This is what the connection should look like:

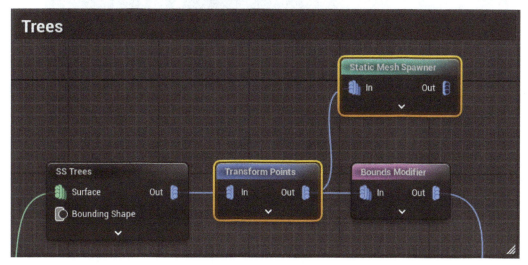

Figure 2.65 – Connecting the Static Mesh Spawner node to Transform Points in the Trees PCG graph

6. The last step is to connect the **Difference** node *output* to the **Static Mesh Spawner** node *input* in the **Rocks** graph network. Note that a new node appears that connects the main nodes. This is the way Unreal translates and converts the connection link between nodes that are different from one another!

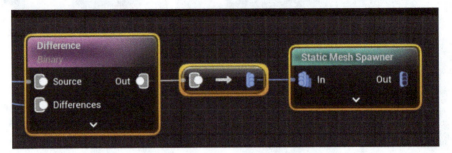

Figure 2.66 – Connecting the Difference node to Static Mesh
Spawner via the To Point connecting node link

We are ready to see the results, so let's test these nodes out by going to the main viewport and running our PCG volume.

Figure 2.67 – Evaluating the distribution outcomes for rocks and trees

Note that the distribution of the rocks and the trees is even in the landscape terrain! This has been a crucial workout exercise for those who seek a perfect way to build a complex vegetation structure. Finally, you should expect the PCG graph to look like this:

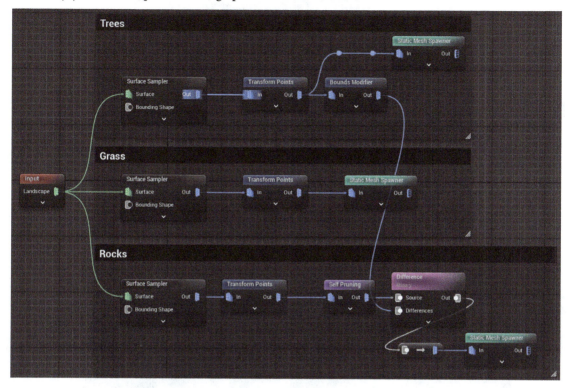

Figure 2.68 – The final layout of the PCG graph network

In the upcoming section, we'll delve into some further options to enhance your PCG graph's performance and how to optimize its PCG volume.

Optimization and performance advice

To wrap up this chapter, it would be great to discuss the importance of optimizing your PCG tool, regardless of whether you use a powerful or lower-end PC or laptop. It's very important to note what to watch out for when you work with the PCG tool to generate dense foliage.

Geometry distribution

It is very important to note that working with photo-scanned geometry (in this case, Megascan trees) can be very performance-heavy for a PCG tool to handle, especially if we decrease the **Point Extents** values inside the Surface Sampler!

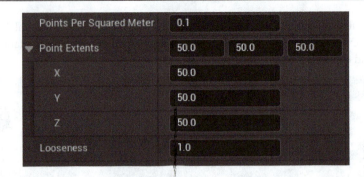

Figure 2.69 – Decreasing the Point Extents values inside the Surface Sampler for the Megascan trees

As an example, I have lowered the **Point Extents** values, and the outcome of the tree distribution looks great but is very performance-heavy for the PCG volume.

Figure 2.70 – Evaluating how spawning trees in close proximity impacts computer performance

The frame rate dropped by *30 fps*! That practice can lower your expectations for a smoother gameplay experience. Enabling **Nanite** can help but only to a certain extent, as you are still putting lots of pressure on your CPU and RAM usage.

Best practice

As explained in the previous section, it is good to increase the **Point Extents** values inside your **Surface Sampler** node and aim for better, more easily optimized static meshes.

Here is a perfect example of spacing the heavy trees in your landscape with those values:

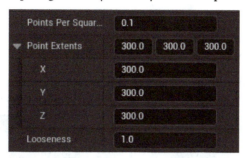

Figure 2.71 – Increasing the Point Extents values to improve the performance of the PCG volume

In *Figure 2.72*, you will see the outcome results of the terrain distribution configuration. This way, you can prioritize which assets you want to distribute more or less on the terrain!

Figure 2.72 – Decreasing the Point Extents values to enhance grass distribution across the landscape

The preceding figure shows the settings that I used for the point distribution of the grass. You can easily see the difference it makes when you try to leverage the amount of static meshes that are spawned on the terrain.

Scaling the PCG volume

This **Transform** settings illustrates the scale values that can be applied to adjust the size of the PCG volume, helping to improve its performance. Essentially, a smaller scale size leads to better efficiency and performance within a scene.

Try to avoid too much scaling of the PCG volume in the viewport, as this can lead to massive performance issues! Make sure to optimize the node settings first before you scale the PCG volume.

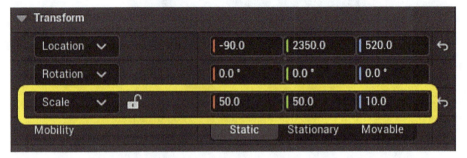

Figure 2.73 – Reducing the size of the PCG volume to improve performance

Unreal Engine 5 takes time to load the PCG volume, which gets scaled up within transformation mode and generates a lot more data to render on the viewport screen. This exerts tremendous pressure on RAM and CPU usage.

Figure 2.74 – Comparing the tree formations and their effects on performance, side by side – X: 50 by Y: 50 units running at 60 fps (left) and X: 100 by Y: 100 units running at 30 fps (right)

In the PCG volume example provided (see *Figure 2.74*), we can compare a smaller and a larger forest size. This comparison sheds light on the importance of scaling and its impact on performance, especially when the PCG volume is tasked with distributing a large number of static meshes at runtime. The objective is to fine-tune the PCG graph inside the PCG volume by adjusting its size and then assessing the performance outcomes.

We've now reached the end of this chapter. You've embarked on the journey of developing your first PCG tool, enabling the procedural generation of a forest!

Summary

In this chapter, you learned how to create your first PCG tool that procedurally generates a forest! You also gained a thorough understanding of the most used nodes to generate different types of foliage on terrain, as well as different nodes, such as **Self Pruning**, **Bounds Modifier**, and **Difference**, to help organize a structure and diversify the selection of spawned static meshes on a landscape.

Additionally, you mastered the techniques to optimize and enhance the performance stability of your designated PCG graph, along with how to implement those strategies to boost the performance of your PCG volume. This insight will help to evaluate the quality of your PCG graphs and improve the efficiency of your procedurally generated system.

In the next chapter, we will learn how to create more complex and fun PCG graphs with the use of Blueprints, delve into further optimizations, and learn the basics of integrating PCG graph components into Blueprints.

Get This Book's PDF Version and Exclusive Extras

Scan the QR code (or go to `packtpub.com/unlock`). Search for this book by name, confirm the edition, and then follow the steps on the page.

Note: Keep your invoice handy. Purchases made directly from Packt don't require an invoice.

3

Building Blueprints with PCG Component

In *Chapter 2*, you learned how to craft your own **procedural content generation** (**PCG**) graph, incorporating various static meshes into the PCG Volume spread across the landscape. Now, in this chapter, we shift our focus to constructing a custom PCG Blueprint and integrating it with the PCG component.

In this section, you'll discover how to create a PCG Blueprint that utilizes the Point Loop function to spawn static meshes. You'll learn to combine different override functions to create unique spawning arrangements of static meshes, ensuring even distribution across the PCG volume.

Initially, you'll engage in a detailed tutorial to master crafting an independent Blueprint, which is a vital skill for those aspiring to develop unique PCG components that are compatible with the level maps inside Unreal Editor. This approach not only equips you with the knowledge of how to construct a custom PCG component, but it also opens up avenues for experimentation.

You'll learn how to create your own PCG Blueprint and implement a PCG Blueprint Element in the PCG graph alongside other nodes. This hands-on experimentation is crucial for a deeper understanding of PCG Blueprints, moving beyond mere adherence to instructions for modifying Blueprint variables.

However, if you prefer a more guided approach, I provide step-by-step instructions to replicate my simple, yet functional, PCG component within the PCG graph. For those interested, my detailed tutorial is available for download via a GitHub link in the *Technical requirements* section of this chapter.

We begin this chapter by diving into the PCG Blueprint, starting with a PCG graph. Then, you'll create your own PCG Blueprint and learn how to use Blueprint node logic for the random transformation of spawned static meshes on the landscape. Next, we'll incorporate this custom Blueprint into the PCG graph, connecting it seamlessly with other nodes.

This chapter delves into several key areas:

- Introduction to programming in a PCG Blueprint
- Executing PCG data setup

- Creating a Point Loop function
- Implementing a custom PCG Blueprint into an Actor Blueprint

To follow along with the PCG Blueprint tutorial, we will utilize a selection of assets that were accumulated in the previous chapter.

Upon completing the tutorial on building a custom PCG Blueprint, it's recommended to revisit this chapter. However, on your second attempt, experiment with your own models and settings. The main goal of this chapter is for you to gain the skills necessary for crafting your unique PCG Blueprints and understanding their underlying techniques. By the end of this chapter, you'll gain hands-on experience in creating your own PCG Blueprint, including designing a simple custom node and integrating it with various nodes and settings within a PCG graph.

We'll begin with an introduction to the PCG Blueprint, covering fundamental aspects of the Blueprint graph, including crucial variables and functions.

Technical requirements

To complete this chapter, you'll need the following:

- A good computer that can run a PCG and those have the following specifications:
  - A high-performance multi-core CPU (AMD Ryzen 7/9, Intel i7/i9)
  - A GPU (NVIDIA RTX, AMD Radeon RX with 8+GB VRAM)
  - At least 16 GB RAM
- Basic knowledge of Blueprint creation in Unreal Engine 5. This is crucial, especially for those new to Blueprint creation. The chapter is designed to be accessible to those with even a fundamental understanding of Blueprint development.

The files for this chapter are placed at Github link. `https://github.com/PacktPublishing/Procedural-Content-Generation-with-Unreal-Engine-5/tree/main/Chapter_3/UE5_PCG_Chapter_03`

In *Chapter 2*, you learned the practical side of creating your own PCG tool with the PCG nodes. Now, you will combine this with the PCG Blueprint, which we will create in this example.

The code files for the chapter can be found at `https://packt.link/PPMwK`

PCG graph preparation

Before we move on to the main topic of the PCG Blueprint, we will prepare and create a PCG graph, which we will use later. This will be used as your placeholder, which you will work with inside the Blueprint later. Follow these steps to create a PCG graph:

1. Create your PC Graph and rename it `PCG_Generator`. Open your graph and let's start adding some nodes, which we will use later.

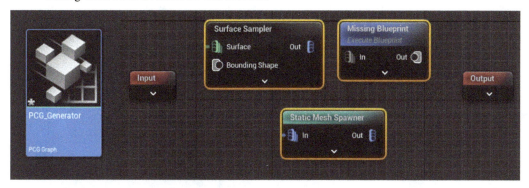

Figure 3.1 – Create a new PCG graph and add those nodes inside your graph

2. As we move on with setting up the nodes, it would be beneficial to explore and explain the purpose of **Execute Blueprint** in this exercise.

Figure 3.2 – The Execute Blueprint node

As you may have noticed, that node looks quite new to you. This is an **Execute Blueprint** node, which we will use to execute our PCG Blueprint later in this chapter. This node enables you to combine custom logic and operations defined in a PCG Blueprint with the PCG graph, providing greater flexibility and control over how procedural elements are generated and manipulated.

This should be enough to start with, and now, we can proceed to the main segment, which involves creating your own PCG Blueprint. This section of the tutorial might be harder and requires additional time to fully grasp. Take your time to understand the concepts and practice at your own pace.

Creating a PCG Blueprint

We ended the previous section by creating your PCG graph, and in the next part of this exercise, in this section, we are going to create a PCG Blueprint. This will be a very exciting exercise to follow and this method is very similar to creating any other Blueprint. Follow these steps:

1. To begin with this exercise, let's get back to **Content Browser | PCG**, and in your folder structure, right-click on the **PCG** folder space and search for **Blueprint Class**.

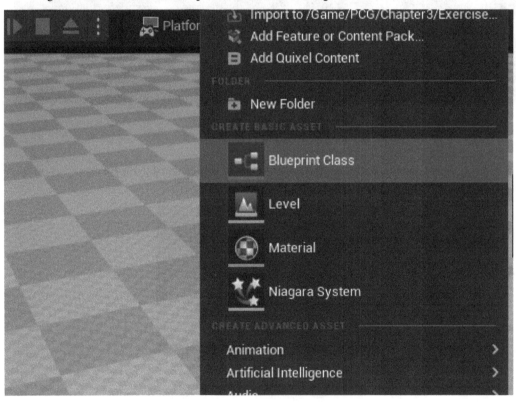

Figure 3.3 – Selecting Blueprint Class

2. With the window open for the **Blueprint Class** selection, in the search bar underneath where it says **ALL CLASSES**, search for `PCGBlueprintElement`. Select it and create `PCGBlueprintElement`! Rename it to `BPPointGenerator`

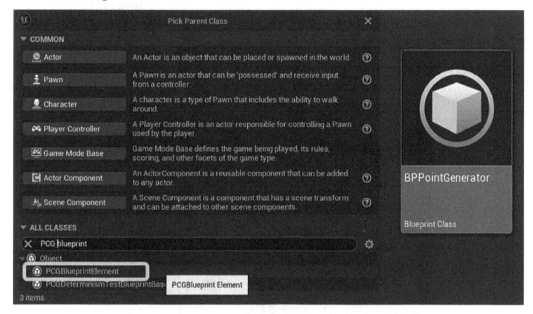

Figure 3.4 – Searching PCGBlueprintElement under ALL CLASSES

Before diving into our Blueprint creation, it's important to understand the distinctions between a typical Blueprint and the one we'll be using for this exercise. While we'll implement a similar methodology to that used in regular Blueprint creation, our focus here will be primarily on creating functions as the main mechanisms for operating elements within the PCG Blueprint.

The Unreal PCG tool offers template functions that are inherent in the Blueprint structure, assisting us in developing the specific functionalities within them.

Building an Override function

In this chapter, we will delve into the capabilities of the Override function. This potent feature enables the modification or replacement of sections within a PCG graph during runtime. The Override function is particularly valuable for executing complex calculations within a node, which are then incorporated into the broader PCG graph. This facilitates dynamic alterations and enhancements to procedural content based on real-time inputs or conditions.

Execute with Context

The core aspect of this exercise revolves around creating the primary logic for point distribution and utilizing it to modify points within the PCG data collection. Previously, we focused on the Point Loop Body function. Now, our task is to develop a new function, which is again located in the **FUNCTIONS** tab, specifically under the **Override** section:

1. Navigate to the **Override** panel inside the **FUNCTIONS** tab and search for `Execute with Context`. Click on **Execute with Context** to create a new function.

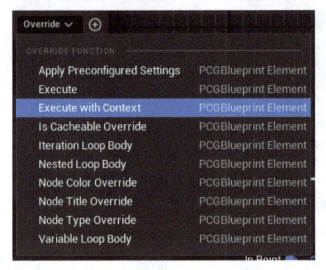

Figure 3.5 – The Override function window

2. As you may have noticed, that has created a new function that is ready to work with the example, and this time, we will take care of the main context and build the structure around it. From **In Context** in **Execute with Context**, drag out the pin and select **Promote to variable**.

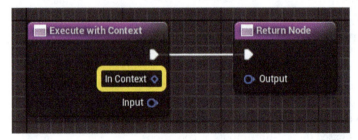

Figure 3.6 – Inside the Execute with Context function

3. While *Figure 3.6* shows a screenshot of the **Execute with Context** function, *Figure 3.7* shows how we can select **Promote to variable**.

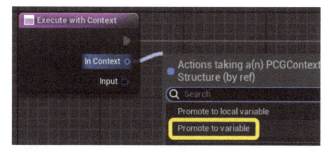

Figure 3.7 – Promote to variable to create the In Context variable

4. In this part, we set the variable to store the context element to ensure that our PCG runs on the same dataset.

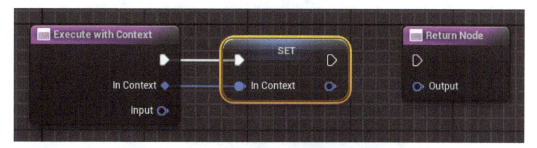

Figure 3.8 – Connecting the In Context variable

5. The next step is to ensure that our input uses PCGDataCollection and here, we will use the **Break PCGDataCollection** function and access the TaggedData input. This will allow us to search and loop through the data points available inside PCGDataCollection.

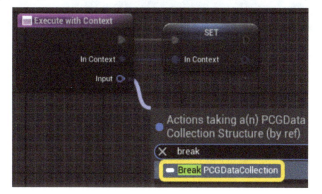

Figure 3.9 – Adding a Break PCGDataCollection node to the graph

6. Next, we will proceed by introducing the **Break PCGDataCollection** node and connecting it to **Input** of the **Execute with the Context** node.

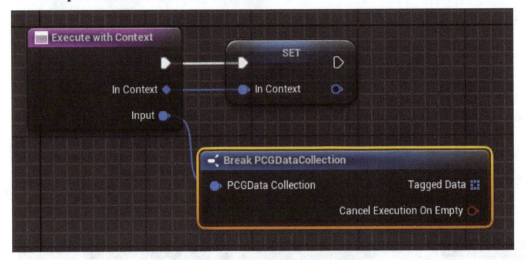

Figure 3.10 – Connecting Break PCGDataCollection

7. The next step is to run the Array Element of `PCGDataCollection` through the **For Each Loop** node. Before doing this, we have to drag the node from the **Tagged Data** input inside the **Break PCGDataCollection** node.

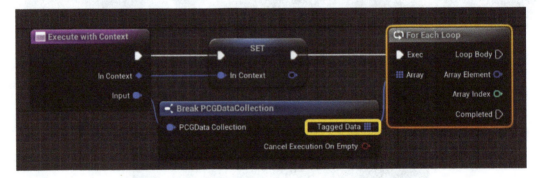

Figure 3.11 – Connecting Tagged Data to the For Each Loop node

> **Note**
>
> `PCGDataCollection` in Unreal Engine 5's PCG framework is a versatile tool for handling a wide range of data types and structures. Its main purpose is to collect, organize, and manage the data needed for PCG, making it easier for developers to create complex and dynamic game worlds.

As we proceed, our next task involves accessing `PCGDataCollection` and dissecting its tagged data. This allows us to retrieve and convert it into `PCGSpatialData`.

`PCGSpatialData` is a potent class filled with extensive information, representing points, volumes, and areas within 3D space. In this specific example, we'll be utilizing it to integrate with level design.

PCG and Point Loop integration

In this section, we'll pick up where we left off in the previous exercise and proceed with integrating the remaining elements into our Blueprint graph. We'll focus on leveraging the data points and expanding them to suit our specific requirements:

1. In the **For Each Loop** node, drag out the connector from **Array Element** and search for the **Break PCGTaggedData** node.

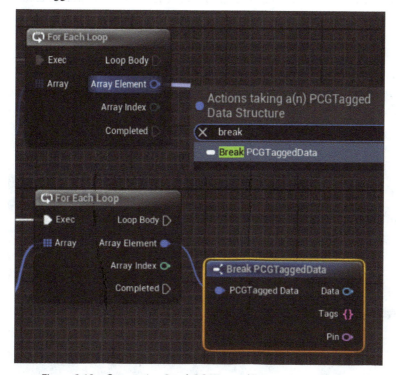

Figure 3.12 – Connecting Break PCGTaggedData to Array Element

That's what your Blueprint should look like at this stage of the tutorial exercise.

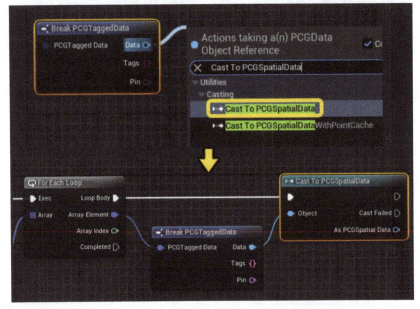

Figure 3.13 – Summary of the linked nodes at this stage

> **Note**
>
> PCGTaggedData is a data structure that holds information related to PCG. This could include various types of data, such as vectors, numbers, or references to other objects or entities within the game world.

2. In the next step, we are going to cast to PCGSpatialData. Under **Break PCGTaggedData**, from the **Data** input, drag out a connector to **Cast To PCGSpatialData**.

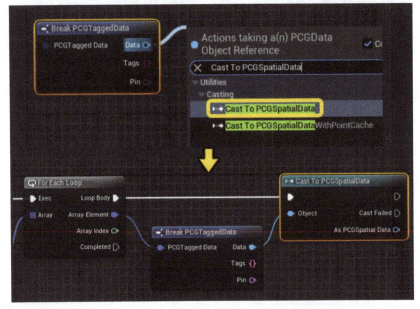

Figure 3.14 – Connecting the Cast To PCGSpatialData node to the graph

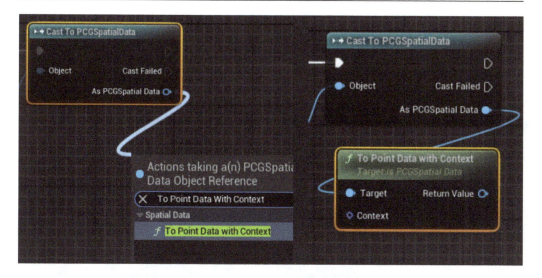

Figure 3.15 – Adding and connecting the To Point Data With Context function

3. Extend the connector to the **To Point Data with Context** function. By connecting to this function, you will be able to link to an existing **In Context** variable, which holds the point data module required for this example.

4. Just as discussed in the previous point, you can now drag the **In Context** variable out and connect it to the **To Point Data with Context** function node.

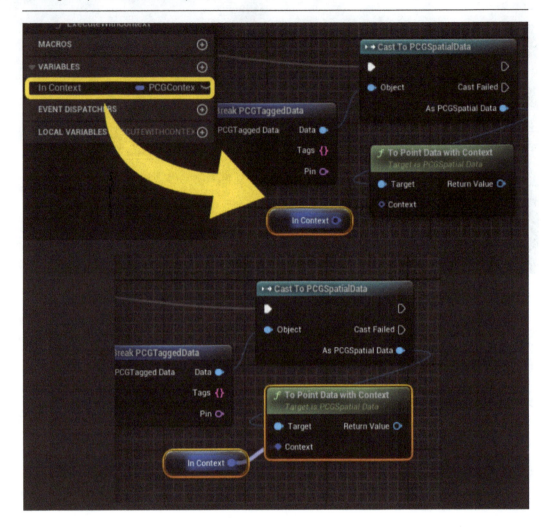

Figure 3.16 – Connecting the In Context variable to the To Point Data with Context node

In this section, we will move on to the Point Loop function.

We are moving on to the next section, where we will search for the Return Node. This will be used to execute and create populated points through a looping action. Drag the **Return Value** input and connect it to the `PointLoop` function.

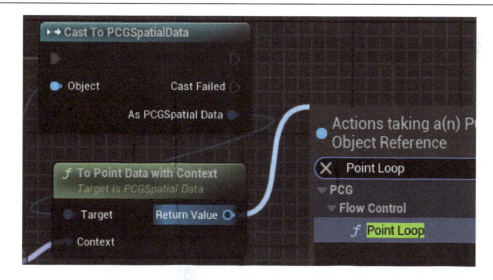

Figure 3.17 – Adding and connecting the Point Loop function

5. In this part, we have connected the PointLoop function to its return value of the **To Point Data with Context** function.

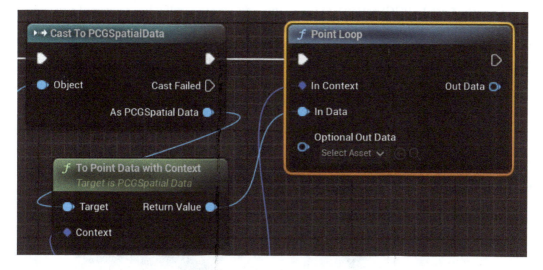

Figure 3.18 – The Point Loop function overview

6. Once you have added the PointLoop function to your graph, it's necessary to connect the **In Context** variable that we created to the PointLoop function. The concept of **In Context** is to provide a current configuration that will adapt to our PCG graph.

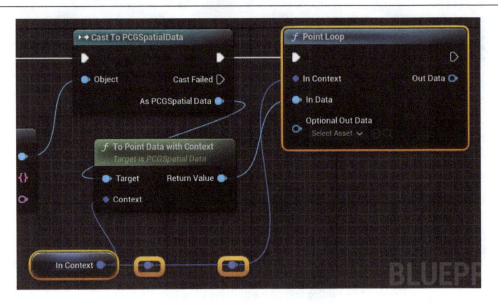

Figure 3.19 – Organizing the graph

7.   Usually, the rule of thumb is to connect the new **Return Node** to the **Cast Failed** output node from the **Cast To PCGSpatialData** node. This way, we can test if it has converted a specific type of object, which, in this case, is PCGTaggedData, from the spatial data to another.

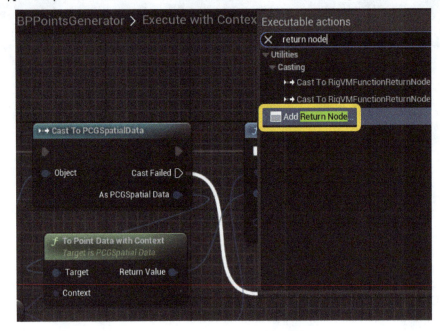

Figure 3.20 – Adding a Return node

Let's add **Return Node** and connect it to the **Cast Failed** output. It will work as a return value if the `PointLoop` function fails to work.

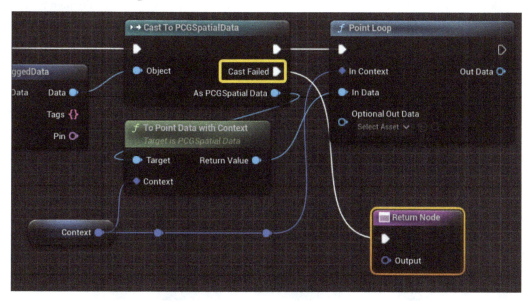

Figure 3.21 – Return Node linked to the Cast Failed input node, which
will return if the PointLoop function fails to work

Let's pause here to reassess the graph. It's important to examine the quality of the existing structure before we continue building the rest of the graph. So far, for the **Execute with Context Override** function, we've connected all the essential nodes leading to the `PointLoop` function. Moving forward, we'll work toward returning the existing `PCGDataCollection` as a **Return Node** output.

In the next part, we will access `PCGTaggedData` and make it work with `PointLoop`:

1. Moving forward, we will begin by accessing `PCGTaggedData` using the **Array Element** from the **Make PCGTaggedData** node, which you need to drag out from the **Out Data** output in the `PointLoop` function.

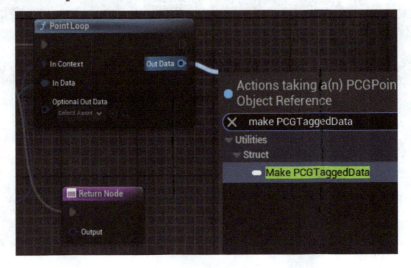

Figure 3.22 – Adding the Make PCGTaggedData node

Let's add the **Make PCGTaggedData** node to our graph and connect it to the **Out Data** output node of `PointLoop`.

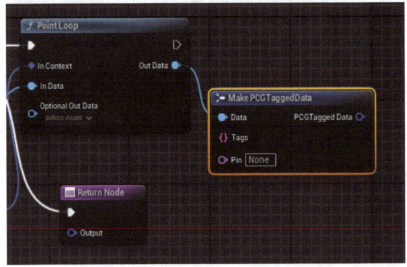

Figure 3.23 – Connecting the Make PCGTaggedData node to the Out Data output of PointLoop

2. After connecting **Make PCGTaggedData** to the **Out Data** output, we can connect the rest of the graph.

3. This is where we are at the moment, and the PCG Blueprint section should look like this:

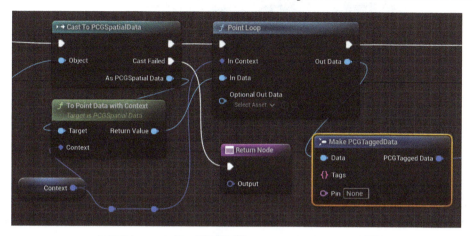

Figure 3.24 – Summary of the linked nodes at this stage

With this set for the **Make PCGTaggedData** node, we need to create an array to access the essential tagged data from this node. Drag out a connector from the PCGTaggedData output, search for Make Array, and connect **Make Array** with the main node.

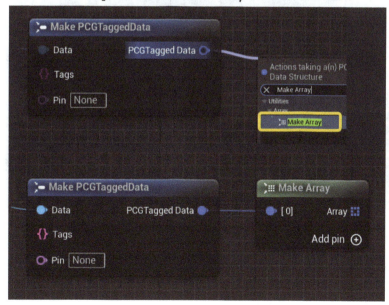

Figure 3.25 – Connecting Make Array to the PCGTaggedData output

4.   The next part is to create a new variable, which will be used as a reference to `TaggedData`, and it will be used to store and utilize data points from the `PCGDataCollection` repository.

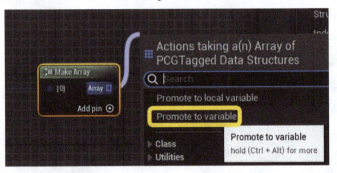

Figure 3.26 – Promoting variable for TaggedData

5.   Change the name of your new variable to `TaggedData`, ensuring it is directly linked to the **Make Array** node. Now, let's proceed by connecting the nodes to further develop your graph.

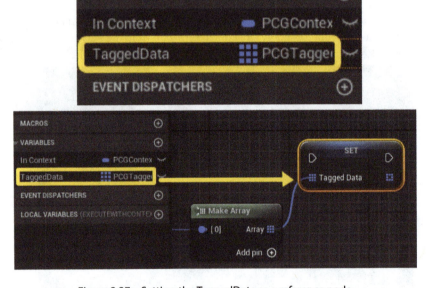

Figure 3.27 – Setting the TaggedData as a reference node

6.   With the `TaggedData` variable set in place, we can now connect it to **Make Array** to form the continuous flow of the PCG Blueprint graph.

7. Right-click on the Blueprint graph space and search for **Make PCGDataCollection**. This will be used to plug the `TaggedData` reference and connect it to the output inside **Return Node**.

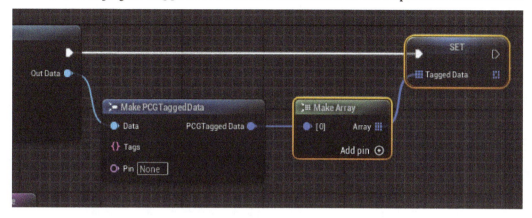

Figure 3.28 – Connecting the TaggedData reference to the PointLoop function

8. Let's drag the pin from the **Tagged Data** variable and search for the **Make PCGDataCollection** node. Make sure that both nodes are connected.

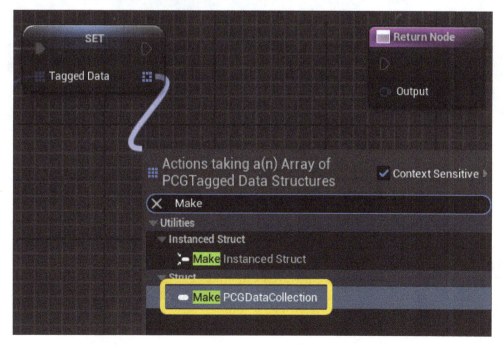

Figure 3.29 – Adding Make PCGDataCollection

9.   We are now proceeding to connect the **Make PCGDataCollection** node to **Output** of **Return Node**, completing the structure of this graph.

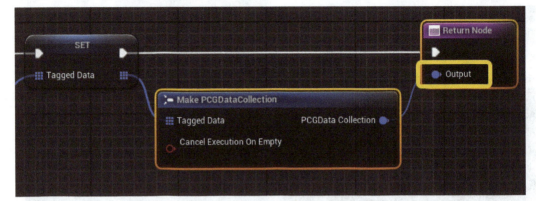

Figure 3.30 – Connecting Make PCGDataCollection to Output on Return Node

Here is the last section of the puzzle. We have connected all the necessary nodes ending at **Return Node**.

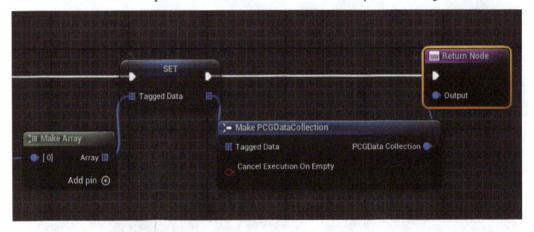

Figure 3.31 – Summary of the linked nodes at this stage

Finally, we have reached the end of creating the **Execution with Context** Blueprint for the PCG Blueprint part. Before we can carry on to the next section, I would like to share with you what you have learned so far and how we can use this exercise to structure an overview of the process, starting from **Execution with Context** and ending with the PointLoop function in a typical Unreal Engine PCG workflow:

- **Execution with Context:**

  - **Description:** This is the starting point of the process. Execution with Context represents the current state or environment in which your PCG logic operates. It might include information about the current level, player data, or other relevant environmental factors.

- **Purpose**: To provide the necessary initial data and conditions for the PCG process.

- **Break PCGDataCollection**:

  - **Description**: `PCGDataCollection` is a structure or collection that holds various PCG data elements. The Break operation deconstructs this collection into its individual components.

  - **Purpose**: To access specific data pieces required for further operations in the PCG process.

- **For Each Loop**:

  - **Description**: This loop iterates over each element in a collection. In the context of PCG, this would typically be a loop over the elements obtained from `PCGDataCollection`.

  - **Purpose**: To process each piece of data individually, which is essential for applying specific PCG rules or modifications.

- **Break PCGTaggedData**:

  - **Description**: `PCGTaggedData` is a part of the collection that usually contains tagged or categorized information. Breaking it down helps in accessing these specific tags and their associated data.

  - **Purpose**: To utilize tagged information for more refined control over the procedural generation.

- **Cast To PCGSpatialData**:

  - **Description**: This is a casting operation where you attempt to treat a general or less specific data type as `PCGSpatialData`, which is more specific and likely related to spatial properties in the PCG context.

  - **Purpose**: To access spatial attributes or properties necessary for the PCG process, such as positions, volumes, or areas in the game world.

- `PointLoop`:

  - **Description**: The `PointLoop` function is likely a part of the PCG process where you iterate over a set of points or spatial data.

  - **Purpose**: To apply specific procedural generation rules or modifications to each point or spatial element, integrating all the previous data manipulations and casting operations.

In the upcoming section, we'll be developing the Point Loop Body Override function. This function is essential for completing `PointLoop` for the `TaggedData`. During this process, we will establish a mathematical formula that enables the static mesh to be distributed randomly across the landscape. This includes varying the position, rotation, and scale of the spawned static mesh!

Point Loop Body

When you open a PCGBlueprint Element, you will find lots of similarities with the typical PCG Blueprint. In this case, the only difference you might encounter is the formation of the graph within. For this, I will show you a step-by-step flow that will be used to create the structure for this specific example:

1. Once you open the PCG Blueprint, on your left-hand side, you will find the tab with **FUNCTIONS** on it. Hover your mouse over that tab and it should show up with **Override**.

Figure 3.32 – Location of the Override functions

2. Click on **Override** and through the list, search Point Loop Body. This will create a new function inside the function menu tab.

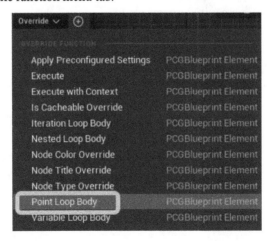

Figure 3.33 – Adding a Point Loop Body function

3. **Point Loop Body** is the initial input node that provides all the access to the context and the data that we need to iterate across all the points in the graph.

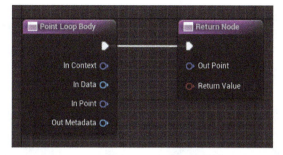

Figure 3.34 – Newly created Point Loop Body Override function

The `PointLoopBody` function will loop through all the points inside the `PCGPoint` structure and it will be used as a logic algorithm to spawn the static meshes through the PCG graph later in a mathematical manner. This process will return the loop function as an **Out Point** node inside **Return Node**.

4. From the **In Point** input, let's search `Set Members in PCGPoint`. This will be the node that accesses the `PCGPoint` library and helps us modify its information, such as Transform and Density. We will talk about this more later in this chapter.

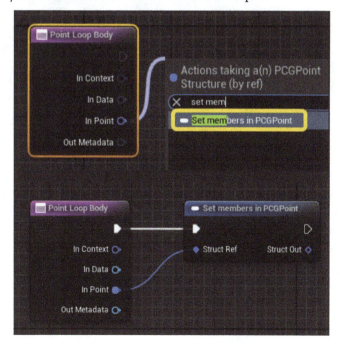

Figure 3.35 – Adding the Set members in PCGPoint node and connecting to In Point

5.   With all those nodes prepared, let's connect all the nodes. Make sure to connect them appropriately just like it is shown in the following graph!

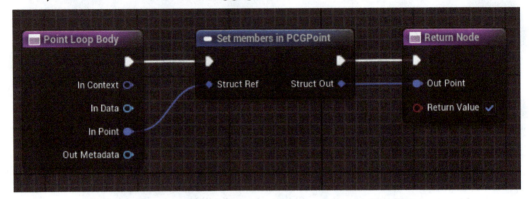

Figure 3.36 – In Point to Struct Ref and Struct Out to Out Point in Return Node

6.   This can be quite easy to forget, but make sure to tick the **Return Value** box! This will enable you to generate the loop pattern later.

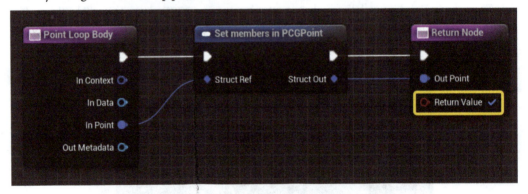

Figure 3.37 – Enable Return Value under Return Node

The next section of the exercise requires you to pay a little attention to detail. We will be engaging with several mathematical equations and linking them to specific nodes. These nodes are integral in coordinating the Transform positions of each spawn point in space. This setup is then utilized to spawn static meshes randomly across the landscape, with each mesh positioned differently.

Transform nodes and random vector variables

To begin, we need to add the nodes that are mainly vectors, and they will be used to spawn objects in space at random positions, rotations, and scales!

1. On the left-hand side panel, under the **VARIABLES** tab, press the + sign button to add variables. Convert them to **Vector** and name them in the following order: `RandomPosition`, `RandomRotation`, and `RandomScale`. Make sure to make them public by clicking on the eye icon.

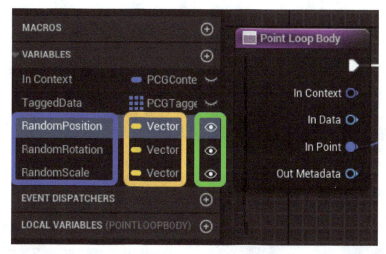

Figure 3.38 – Adding the vector variables to Point Loop Body

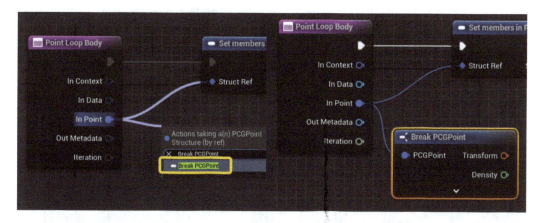

Figure 3.39 – Adding and connecting the Break PCGPoint node

2. From the **Point Loop Body** input, drag out the connector from the **In Point** output and add the **Break PCGPoint** node.

3. As you can see, there is a **Density** input node added to the **Break PCGPoint** node, along with the **Transform** input. We can disable the **Density** input as we don't need it for this exercise because we are not going to change the volume number of points in the PCG graph. To switch

it off, select the **Break PCGPoint** node, and on the right-hand side, under the **Details** panel, deselect all of the properties apart from the **Transform** input!

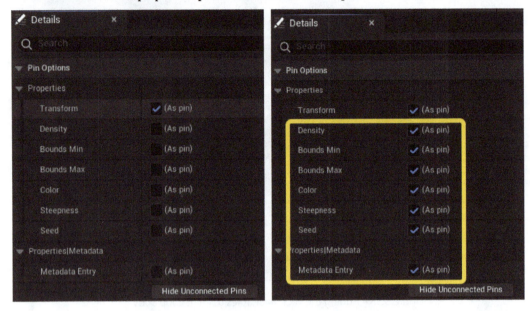

Figure 3.40 – Disabling properties and exposing a Transform property value

4. With that part out of the way, we can now focus on building the Transform nodes and hooking up a newly created vector variable to the equation! Drag out the pin from the **Transform** input under **Break PCGPoint** and search `Break Transform`.

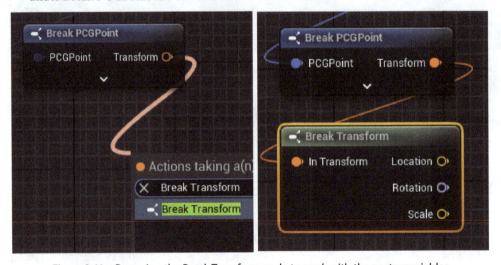

Figure 3.41 – Preparing the Break Transform node to work with the vector variables

Drag and drop the get `RandomPosition` vector from the **VARIABLES** tab.

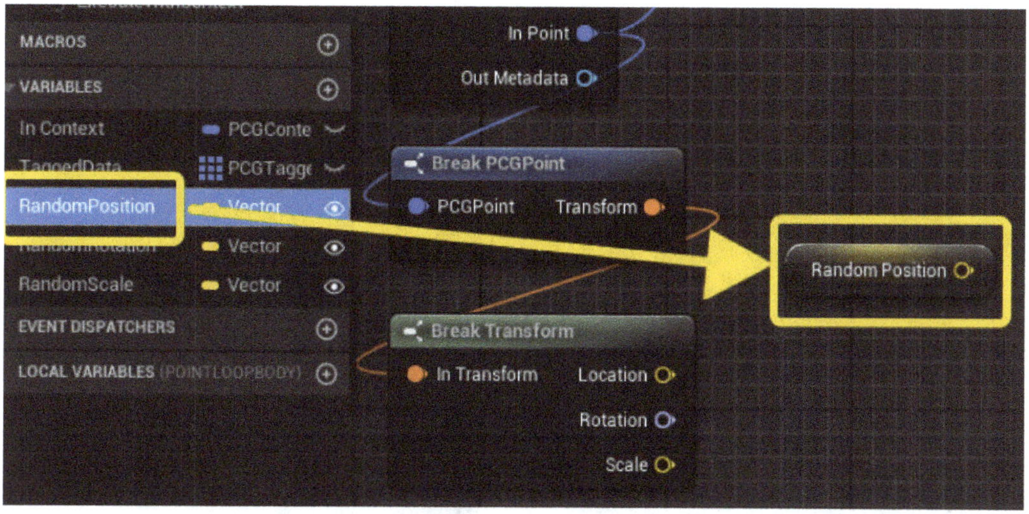

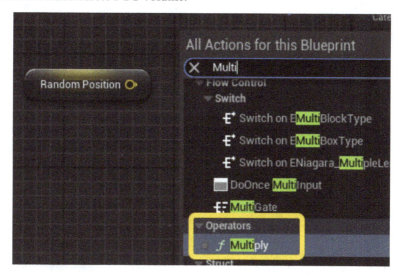

Figure 3.42 – Drag and drop the RandomPosition vector variable from the VARIABLES tab

5. Right-click on the Blueprint space and search for the `Multiply` operator. Multiply it together with the location so that the location will be used to uniformly translate the static meshes from the world location across **PCG Volume**.

Figure 3.43 – Adding the Multiply node to the graph

6.   Connect it all to the **Break Transform** node.

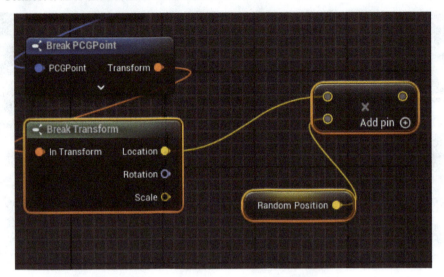

Figure 3.44 – Multiply the Random Position node with the Break Transform location node

7.   Another step involves introducing an additional node to our setup, specifically a Make Transform node. This node will facilitate the consolidation of the variables we've employed in this example, eventually integrating them into a single, functional algorithm.

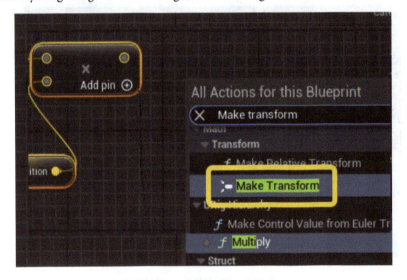

Figure 3.45 – Adding Make Transform

With the **Make Transform** node selected, connect the node to the `Multiply` operator of the **Random Position** vector.

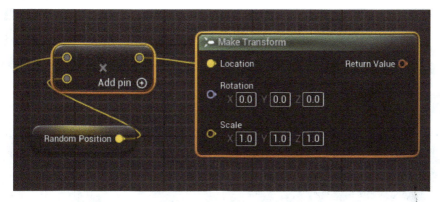

Figure 3.46 – Connecting with the Make Transform node location

Next, let's focus on the Set Members in PCGPoint node, which we introduced in the first part of the tutorial. It's important to activate the Transform pin within that node. While it may seem familiar, this is the crucial node that needs to be linked back so that the data from the `PCGPoint` will function effectively within the Point Loop Body Override function. Let's follow these steps to do the same:

1. With the **Set members in PCGPoint** node selected, hover with your mouse to the right-hand side of the screen to the **Details** panel. Enable the **Transform** property and connect the **Make Transform** node to **Set members in PCGPoint** via the **Transform** property.

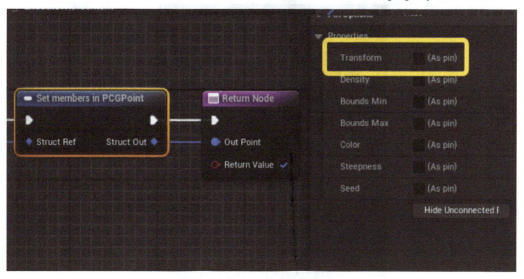

Figure 3.47 – Adding the Transform node to the inside of Set members in PCGPoint

2. In this part, we will connect the Make Transform node to the Set members in PCGPoint Transform node.

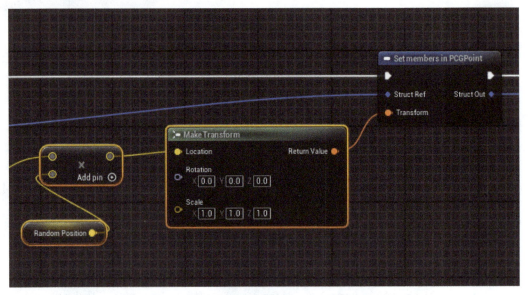

Figure 3.48 – Connecting Make Transform with the Set member in PCGPoint Transform node

3. Inside the **Make Transform** node, right-click on **Rotation** and then click **Split Struct Pin**. The reason behind this is to only use X and Y values for this example.

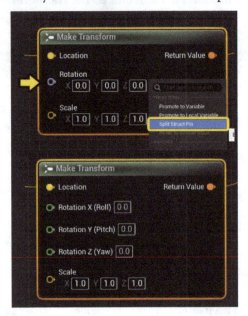

Figure 3.49 – Split Struct Pin to expose the Rotation float values

4. Right-click on the graph and search for the **Random Float** and **Random Float in Range** nodes. Link these nodes to the **Random Rotation** variable in a similar manner to your previous connection with the **Random Position** variable. See *Figure 3.50* as a guide!

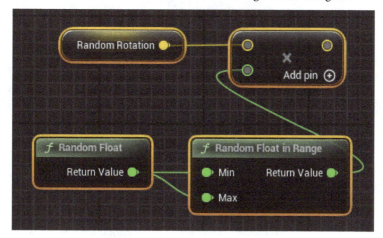

Figure 3.50 – Creating a formula for Random Rotation

5. Do the same but with the **Random Scale** variable.

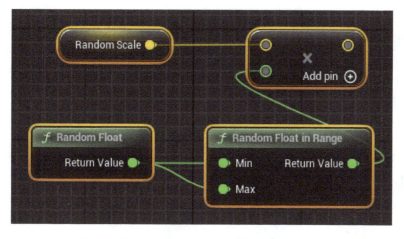

Figure 3.51 – Creating a formula for Random Scaling

6. For the **Random Rotation** structure, right-click and search for the **Break Vector** node, and then connect it just like in *Figure 3.52*.

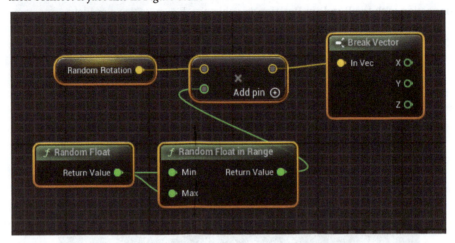

Figure 3.52 – Adding and connecting Break Vector to the Random Rotation equation

7. Now, you can connect the **Random Rotation** equation to the **Make Transform** node for X and Y values.

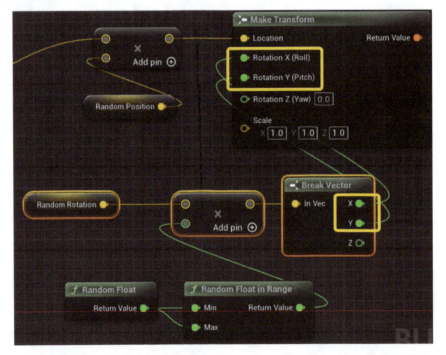

Figure 3.53 – Connecting the Random Rotation nodes to Make Transform via Break Vector

8. The last part is to connect the **Random Scale** equation to the **Make Transform** node. In this case, we don't need to break anything because we want our scale to act uniformly.

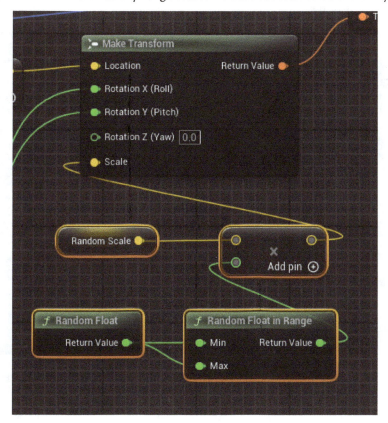

Figure 3.54 – Connecting the Random Scale node to the Make Transform Scale node

Your **Point Loop Body** graph in the PCG Blueprint should now be fully assembled with nodes resembling those in the following figure. Remember to compile and save your PCG Blueprint once everything is set up correctly!

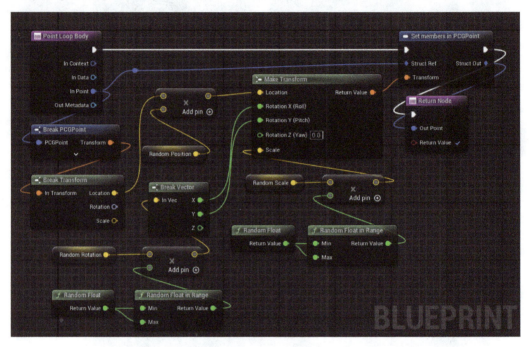

Figure 3.55 – Overview of the Point Loop Override function and how it is all connected

You've successfully completed your custom PCG Blueprint, gaining skills in creating personalized PCG Blueprints for procedural generation content. Next in this tutorial, we'll concentrate on developing the Actor Blueprint and integrating the PCG component. Following that, we'll revisit the PCG graph to conduct tests with your PCG Blueprint.

Building an Actor Blueprint

Now, we've reached the stage where we can put the Blueprint you developed earlier into action. But before that, we need to add and construct an Actor Blueprint. In this Actor Blueprint, we will integrate the PCG component. To start this process, we'll first add the Actor Blueprint to the content browser's folder and kick off the PCG experience:

1. Inside the folder where you store the PCG Blueprint, right-click, go to **Blueprint Class**, and search for a new Actor Blueprint. Let's call it `BP_MeshGenerator`.

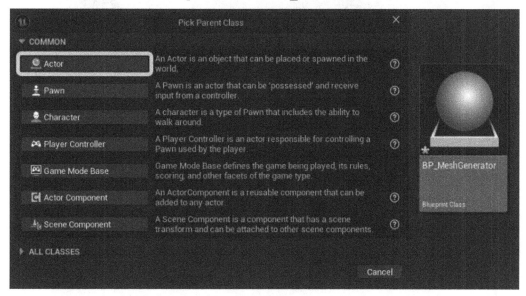

Figure 3.56 – Creating a new Actor Blueprint and renaming it BP_MeshGenerator

2. Open your new Blueprint and start adding variables for this exercise. All of the variables should be vectors. Let's rename the vectors with the following names: `RandomPosition`, `RandomRotation`, and `RandomScale`. Make sure all the parameters are set to public by clicking on the closed eye icon.

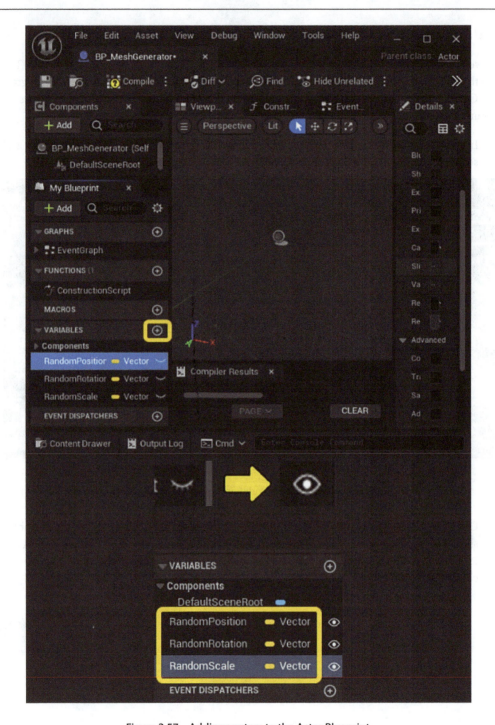

Figure 3.57 – Adding vectors to the Actor Blueprint

3. After setting everything up, the following step involves adding a **Box Collision** component within the **Components** window. The PCG component works very intelligently, utilizing any collisions as a PCG Volume space. Box Collision components are used to populate static meshes within these spaces!

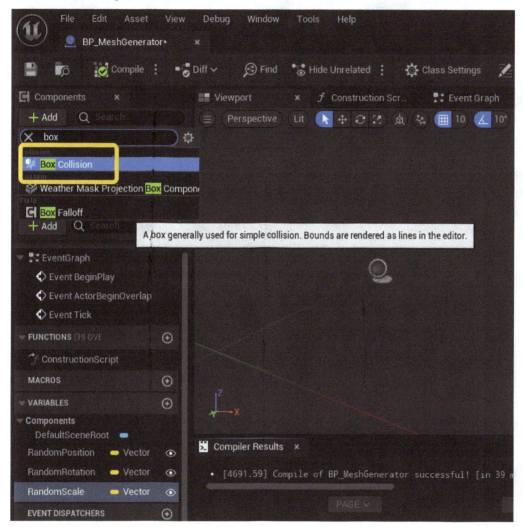

Figure 3.58 – Adding a Box Collision component inside the Actor Blueprint

4. We are almost done with the Actor Blueprint, but to make our results more visible, we need to increase the **Box Collision** component's **Box Extent** scale to 2000.0 for the X value, 2000.0 for the Y value, and 100.0 for the Z value!

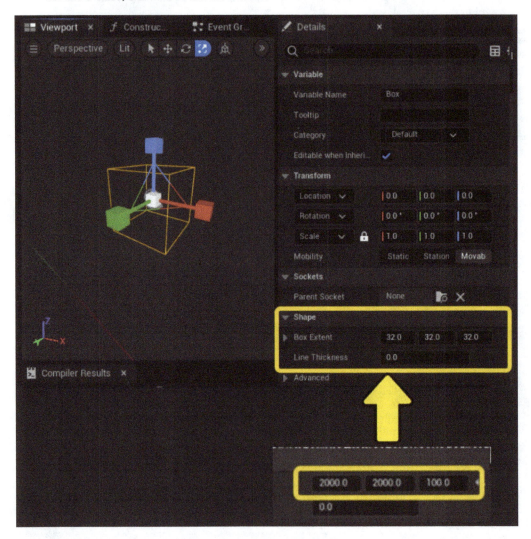

Figure 3.59 – Resizing Box Extent of the Box Collision component inside the Actor Blueprint

The Box Collision component must be quite large to fit in all the static meshes within its area, and it can be made so with the help of the PCG graph component embedded inside the Actor Blueprint.

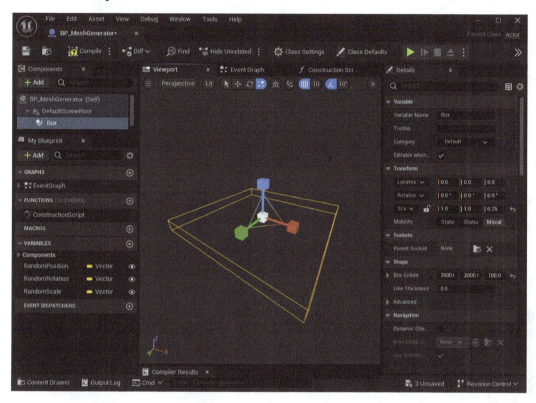

Figure 3.60 – Overview of the Box Collision component after adding a few modifications

5. Last but not least is adding the PCG component to this Actor Blueprint. Click on the **Add** button and search for PCG.

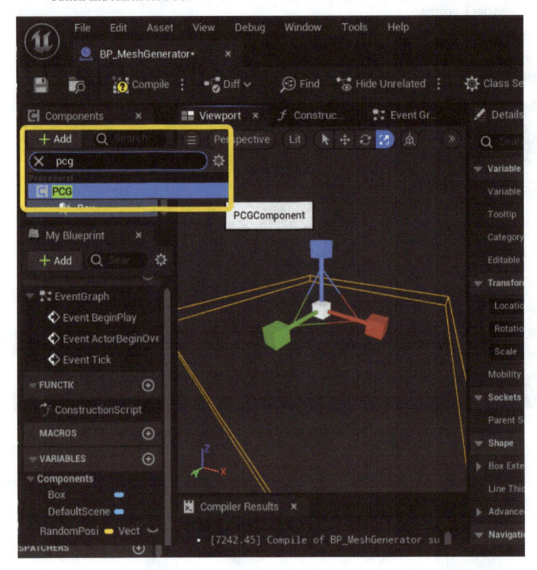

Figure 3.61 – Adding a PCG component inside the Actor Blueprint

6.   With the **PCG** component selected, search for the PCG graph, the one we created at the beginning of this chapter, called PCG_Generator. After this, you can close your Actor Blueprint!

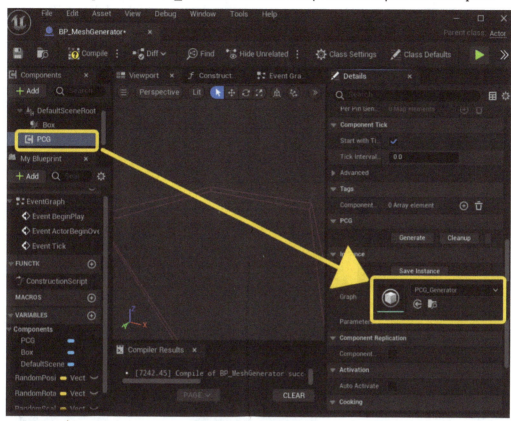

Figure 3.62 – Adding a PCG_Generator graph to the PCG component

You just learned how to prepare another Blueprint that can accommodate an existing PCG Blueprint. The Actor Blueprint will work as a placeholder to store the PCG component that you can modify while working on your projects. Moving forward, let's compile your Actor Blueprint and close it. Drag and drop your BP_MeshGenerator Actor Blueprint into the scene.

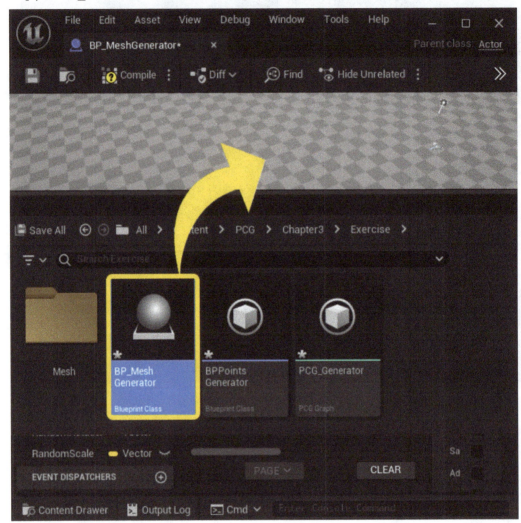

Figure 3.63 – Drag and drop the BP_MeshGenerator Actor Blueprint into the scene

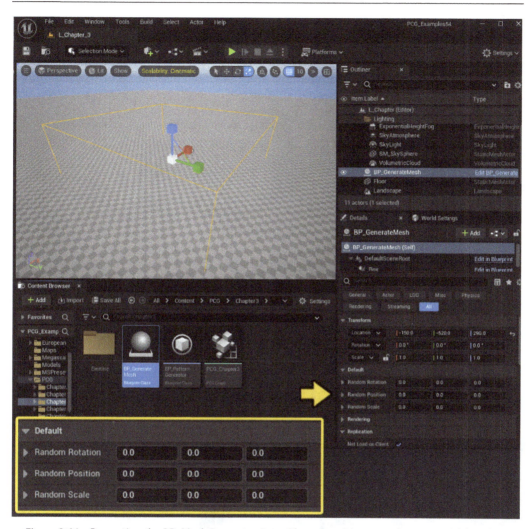

Figure 3.64 – Presenting the BP_MeshGenerator Actor Blueprint with exposed vector variable values

At this stage, nothing will spawn as we haven't set up any nodes in the PCG graph yet. We'll address this in the next step, but for now, let's keep the BP_MeshGenerator Actor Blueprint in the scene. You might have observed that your Blueprint contains exposed vector variables, which you created earlier during the Actor Blueprint tutorial. These are going to be useful later in our exercise!

Next, we'll integrate the PCG Blueprint within the PCG graph, ensuring the necessary variables are accessible for controlling the BP_MeshGenerator Actor Blueprint on the landscape.

Connecting nodes in a PCG graph

In this final section of the chapter, we will dissect the nodes required to construct the graph and introduce our PCG Blueprint into the graph. Let's begin this exercise by opening the PCG_Generator PCG graph and adding a few nodes inside it:

1. Inside your PCG graph, right-click and search for the following node: **Get Actor Data**.

Figure 3.65 – Adding Get Actor Data inside the PCG graph

2. Right-click inside the graph and search for **Get Actor Property**. Duplicate this node two times so we have three of these nodes in total!

Figure 3.66 – Adding Get Actor Property three times inside the PCG graph

> **Note**
>
> This set of nodes will represent the variables that are already created within the PCG and Actor Blueprints. This will configure the PCG graph to expose and establish connections between these variables and the variables inside your Blueprints.

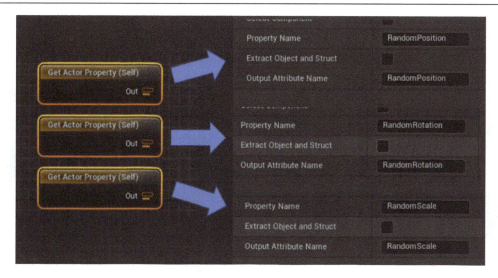

Figure 3.67 – Adding Property Names for each Get Actor Property

3. With each node selected, go to the **Settings** panel and change their **Property Name** and **Output Attribute Name** fields to match your vector variables; these are RandomPosition, RandomRotation, and RandomScale.

> **Note**
>
> Make sure the names of the variables under each **Get Actor Property** node match the exact name of your vector variables!

4. Before connecting the Input node directly to the Surface sampler, an input pin must be added to enable mesh generation on the landscape surface. To do this, navigate to the Settings panel located on the right side after selecting the Input node. Add a new index array, set its **Allowed Type** to **Surface**, and rename it to Landscape. This will display the surface landscape as a green input pin on the **Input** node.

Figure 3.68 – Adding input pin for enabling mesh generation on the landscape surface

5. Let's connect our graph in this manner, just as in the following figure.

Figure 3.69 – View of the PCG graph structure

6. Select the **Execute Blueprint** node and on the right-hand side, under **Blueprint Element Type**, search for your `BPPointsGenerator` PCG Blueprint.

Figure 3.70 – Adding a BPPoints Generator PCG Blueprint inside the Execute Blueprint node

7. With your Blueprint node selected, click on the bottom arrow to unhide the elements that we need to connect the **Get Actor Property** nodes later. This will expose your vector variables.

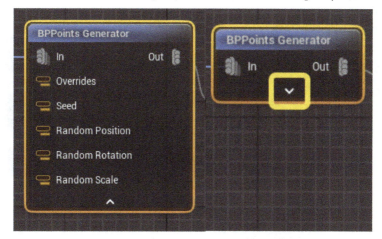

Figure 3.71 – Revealing the nodes

8. The next phase is to connect the Get Actor Property nodes accordingly to each input inside the BPPoints Generator node.

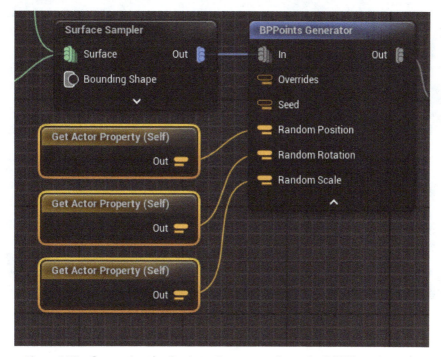

Figure 3.72 – Connecting the Get Actor Property nodes to the PCG Blueprint node

9.   Last but not least is to add the mesh to the **Static Mesh Spawner** node and for this, we will use the S_Mossy_Rock_Assembly_p1FhY_lod3 geometry; you can find this asset inside Quixel Bridge.

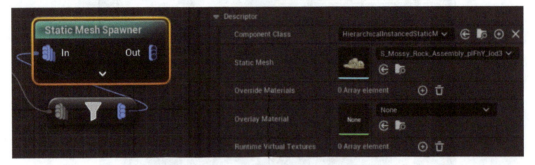

Figure 3.73 – Adding a S_Mossy_Rock_Assembly_plFhY_lod3 static mesh to Static Mesh Spawner

Here is what your final graph may look like with all the nodes connected inside your PCG graph.

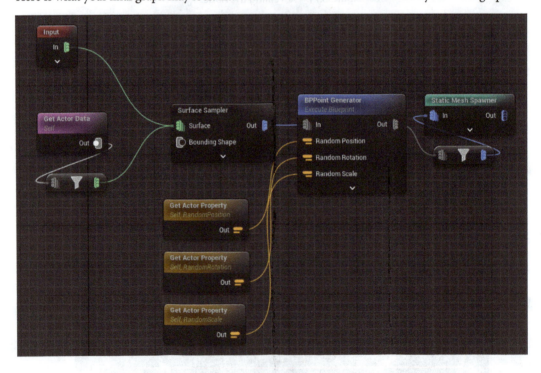

Figure 3.74 – An overview of the PCG graph with all the nodes connected

The final step of this exercise is to test the PCG Blueprint and see what outcomes this will bring. With BP_MeshGenerator selected, you can give it a go and manipulate the values on your right-hand side to change the position, rotation, and scale of your static mesh objects! This ultimately demonstrates the power that you can have over the control of your custom-made PCG component!

Figure 3.75 – The result of the rocks being generated randomly across the landscape

Summary

In this chapter, you've learned about the process of creating a custom PCG Blueprint, utilizing the Actor Blueprint and PCG graph. This is a crucial lesson, as it equips you with the skills necessary to craft your unique PCG graph. This ability is vital for controlling and shaping the content generation over landscapes to your preference. In the upcoming chapter, we will dive into improving the current blueprint, focusing specifically on developing an efficient and more interactive foliage tool.

4

Developing and Optimizing the Procedural Content Generation Tool

In *Chapter 3*, you reached a crucial milestone in the development of the **procedural content generation** (**PCG**) tool by mastering the creation of your own PCG Blueprint. This expertise marks a significant advancement in customizing and enhancing your tool, particularly when the existing PCG library lacks nodes that meet your specific needs.

This chapter will build upon the foundations laid in the previous one but with a focus on the Actor Blueprint and on refining your PCG Actor Blueprint by enhancing the PCG graph with additional nodes. This progression will enable you to tailor and optimize your PCG tool more effectively to your requirements. We will start by introducing additional variables within the Actor Blueprint, which will be essential for the PCG graph later in this chapter.

At the outset, you will focus on enhancing your existing PCG graph, introducing new features and capabilities to the tool in tandem with an Actor Blueprint. This approach is designed to expand the functionality and versatility of your PCG tool.

You will acquire the skill to select appropriate variables that facilitate direct communication with the Blueprints. This knowledge will be advantageous for implementing more interactive modifications directly within the level inside Unreal Engine.

This chapter will cover the following topics:

- Integrating additional variables into the Actor Blueprint
- Refining the PCG graph
- Optimizing the PCG tool's efficiency

To follow along with the PCG Blueprint tutorial, we will utilize a selection of assets that were accumulated in the previous chapter.

We will start by building upon the work from the previous chapter, initially focusing on the Actor Blueprint and then progressing to the upper level of the PCG graph.

Technical requirements

To complete this chapter, you'll need the following:

- A good computer that can run a PCG:

  - Use a high-performance multi-core CPU (AMD Ryzen 7/9, Intel i7/i9) and GPU (NVIDIA RTX, AMD Radeon RX with 8+GB VRAM), with at least 16 GB RAM.

- Basic knowledge of Blueprint creation in Unreal Engine 5. The chapter is designed to be accessible to those with even a fundamental understanding of Blueprint development.

The code files for this chapter are placed at `https://github.com/PacktPublishing/Procedural-Content-Generation-with-Unreal-Engine-5`.

The code in action video for the chapter can be found at `https://packt.link/ChXoU`

Getting familiar with models

In this chapter, we will utilize the existing assets from the `Megascan` library. We'll choose the grass model from the `Content | Megascan` folder:

1. Navigate to `STF | Pack03-LandscapePro | Environment | Foliage | Grass` and search for the grass model.

Figure 4.1 – Locating the SM_PlantGroup01 static mesh

2. The next step of this part is to turn on the **Nanite** mode for each mesh. With those models selected, right-click on one of the models and tick the **Nanite** mode at the top of the menu.

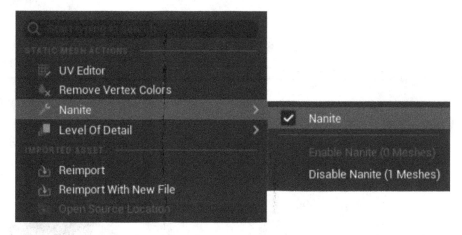

Figure 4.2 – Enabling the Nanite support

We are done with this section; let's move on to more practical fun. In the next section, we will be adding the variables inside the Actor Blueprint and we will be setting up variables and Box collision component.

Adding variables inside the Actor Blueprint

In order to continue with the PCG graph and its modifications, we need to add a few more variables inside the Actor Blueprint, which will be used to communicate directly within the PCG graph nodes.

Setting up variables and the Box collision component

In this part, you will add a set of variables, and the most important is the functionality of resizing the Box collision component, which will be an essential part of scaling up and down the size of your PCG:

1. Open your Actor Blueprint (BP_MeshGenerator), which we created in the previous chapter, and let's start adding new variables:

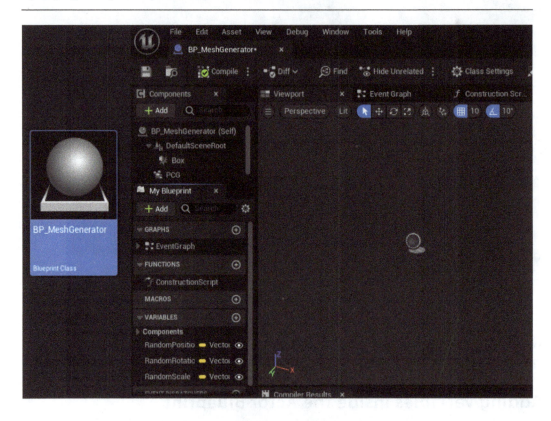

Figure 4.3 – Accessing BP_MeshGenerator

2.   Click on the plus icon next to the **VARIABLES** section and start adding new variables, as shown in *Figure 4.4*. Make sure to make the **Box Extent** vector variable public:

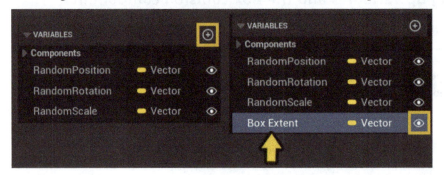

Figure 4.4 – Adding a new variable and setting its visibility to public

> **Note**
>
> Remember to always set each variable to public. This is crucial as it allows you to expose the parameters, enabling you to adjust your PCG tool dynamically through the outliner's details panel, rather than repeatedly returning to the Actor Blueprint for modifications.

Construction Script is a useful feature to have in order to test the size and shape of your mesh generator. Here, you will add a function inside the Construction Script graph. Right-click and search `Set Box Extent function` from the Box component.

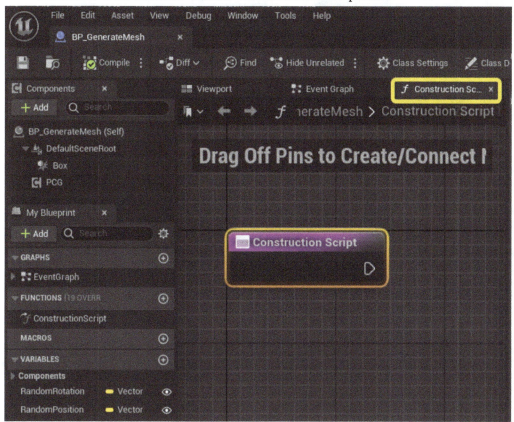

Figure 4.5 – The Construction Script editor graph menu

> **Note**
>
> For those new to the Construction Script, it is a feature designed to create custom logic that automatically runs each time a Blueprint instance is either created or modified in the editor.

3.   Add the selected **Set Box Extent (Box)** function into the **Construction Script** graph.

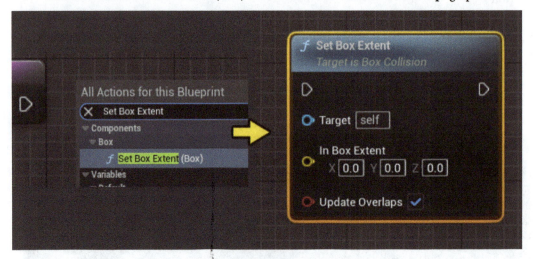

Figure 4.6 – Adding the Set Box Extent function to the Construction Script graph

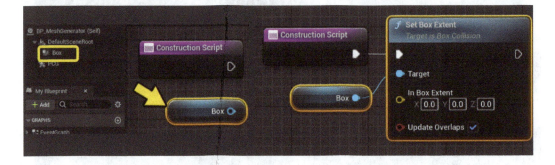

Figure 4.7 – Adding and connecting the Box component to the graph

4.   Before we connect all the dots, let's drag out the **Box** collision component from the **Components** tab. Connect the **Box** component to the **Set Box Extent** function and then connect **Construction Script** to the same function.

5.   Drag the **Box Extent** vector variable out from the **VARIABLES** section and connect it to the **Set Box Extent** target input:

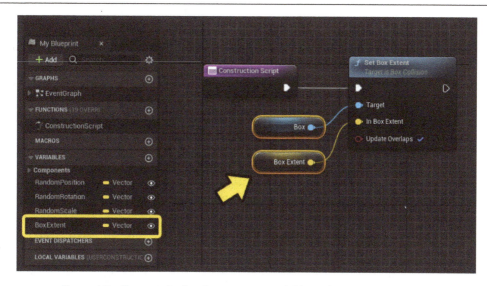

Figure 4.8 – Connect the Box Extent vector variable to the In Box Extent input

6. It is worth testing your tool and checking the scaling property for this Actor Blueprint. Let's hit the **Compile** button (*Figure 4.9*) and you can now close your BP_MeshGenerator Blueprint.

Figure 4.9 – Testing BP_MeshGenerator inside the level editor

7. Don't forget to set your **Box Extent** variable to be visible! You can adjust the size of the volume within the editor panel by utilizing the scale property, and these provide you with two methods for enlarging your PCG volume:

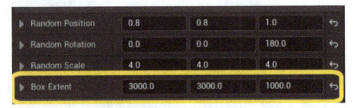

Figure 4.10 – Changing the dimensions in the Box Extent settings panel

You have now completed this section of adding variables inside the Actor Blueprint. Next, we'll explore how to modify and add nodes to the PCG graph you constructed in the previous chapter.

Improving and modifying the PCG graph

In this section, we'll dive into the PCG graph and focus on adding and tweaking several nodes to improve the overall efficiency of the PCG tool.

For this exercise, we will add the `Pack03-LandscapePro` asset, and this time, it will be a grass asset that is procedurally spawned on top of the rocks:

1. To find this asset, go down to `UE5_PCG_Chapter_04 | STF | Pack03-LandscapePro | Environment | Foliage | Grass` and search for `SM_PlantGroup01`.

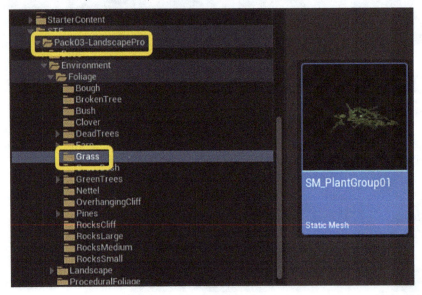

Figure 4.11 – Locating the SM_PlantGroup01 grass foliage asset inside the Grass folder

2. Everything is ready to work with our example and you can close the asset menu for the plant static mesh. Let's kick things off by opening your PCG_Generator graph in the content browser. We'll examine its structure and begin incorporating the necessary nodes for this exercise.

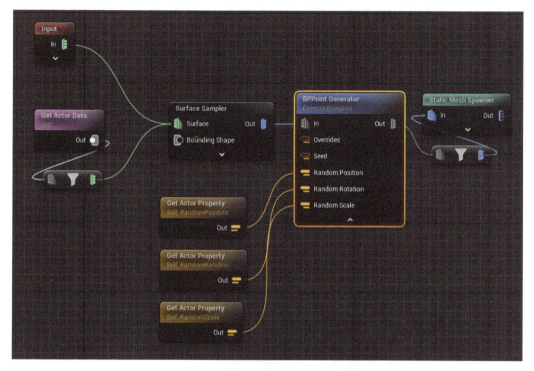

Figure 4.12 – Opening the PCG_Generator PCG graph

3. With PCG_Generator open, check that you have the same graph that we worked on in the previous chapter:

Final adjustments to the PCG graph

The following steps involve linking the nodes from **BPPoints Generator** to the newly created node structure:

1. You will need to incorporate additional nodes that manage the noise and adjust its density together with the **Transform Points**. These adjustments are crucial for controlling the quantity of rocks that are spawned on the landscape and around the rocks. Then, we will finally add grass plants to decorate the environment further. At the end of your PCG graph, disconnect **Static Mesh Spawner** from the **BPPoints Generator** node.

Figure 4.13 – Disconnecting the Static Mesh Spawner node from BBPoints Generator

2.   On the PCG graph space, let's add the **Attribute Noise** node to your graph. **Attribute Noise** is a feature or node that introduces randomness or variation to certain attributes of generated content. This can be useful for creating more natural and varied procedural environments.

Figure 4.14 – Adding the Attribute Noise node to the graph

3.   With **Attribute Noise** selected, on the right-hand side, under the **Settings** panel, press the + icon and change the input source from `@Last` to **Density**.

Figure 4.15 – Changing Input Source to Density

4. With the **Attribute Noise** node selected, navigate to the right-hand side panel and change the **Noise Min** and **Noise Max** values respectively to 0.2 and 1.0. This arrangement will help to regulate the spread of procedurally spawned rocks. Let's clamp those values by ticking the **Clamp Result** box.

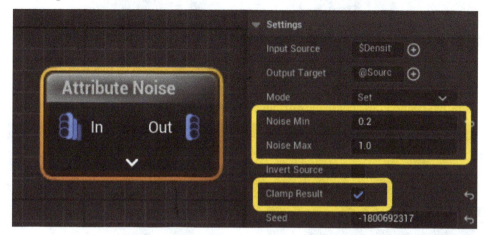

Figure 4.16 – Editing the Attribute Noise settings

5. Let's connect the **BPPoints Generator** and **Attribute Noise** nodes.

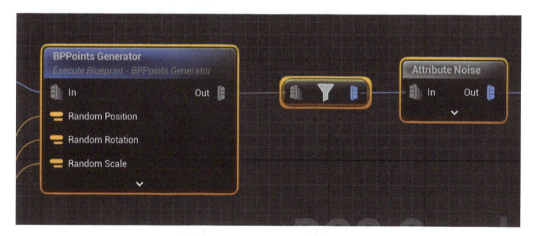

Figure 4.17 – Connecting the Attribute Noise node to the BPPoints Generator node

6. Enhance the noise variation of the spawned objects by integrating a **Density Filter** node. This approach allows for more precise control of the noise boundaries, with values ranging from 0 to 1, compared to the adjustments we made in the **Attribute Noise** node. However, this method offers greater precision and it will remove most of the points that we don't want to use around the radius. Set the **Lower Bound** value to 0.3 for this setup.

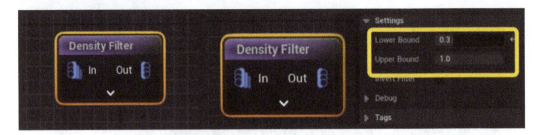

Figure 4.18 – Editing the Lower Bound and the Upper Bound settings for Density Filter

7. Now, let's connect this node to the **Attribute Noise** node to form a continuous PCG graph chain.

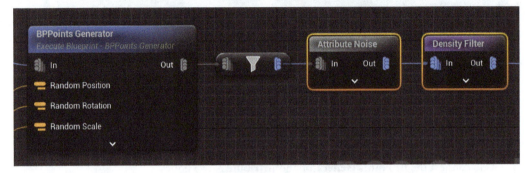

Figure 4.19 – Connecting the Attribute Noise and the Density Filter nodes to the BPPointsGenerator node

8. Right-click on the PCG graph and search for the **Normal To Density** node, which will convert normal vector information into density values.

Figure 4.20 – Adding the Normal To Density node to the PCG graph

9. Maintain the same default values for the settings. The only change needed is to switch the **Density Mode** setting from **Set** to **Add**. This change enables the impact of normal vectors on density, allowing for the adjustment of how different axes influence this effect.

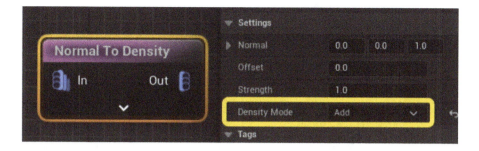

Figure 4.21 – Setting the Normal To Density node's Density Mode to Add

Figure 4.22 – Adding the Transform Points node to the PCG graph

10. This time, it's beneficial to organize the placement of points on static meshes considering the nodes we've previously added. Let's include the **Transform Points** node in your PCG setup!

11. With the **Transform Points** node selected, let's configure a little bit and add a few values to randomize the Transform attributes. Enable the **Absolute Rotation** functionality by pressing the tick box.

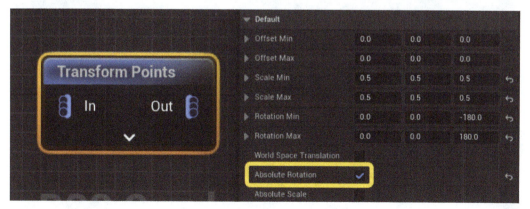

Figure 4.23 – Enabling the Absolute Rotation and editing settings
for each parameter inside the Transform Points node

For better readability, I copied the Transform Points values into this table:

Transform Points	X	Y	Z
Offset Min	0.0	0.0	0.0
Offset Max	0.0	0.0	0.0
Scale Min	0.5	0.5	0.5
Scale Max	0.5	0.5	0.5
Rotation Min	0.0	0.0	-180.0
Rotation Max	0.0	0.0	180.0

Table 4.1- Transform Points values

> **Note**
>
> As this is simply for tutorial purposes, feel free to join your own values to the Transform Points node according to your preference and experiment with it on your own terms in your spare time.

12. Last but not least is to reattach the **Static Mesh Spawner** node. Let's reconnect it to your graph.

Figure 4.24 – Connecting Transform Points in between the Normal
To Density node and the Static Mesh Spawner node

13. With all the nodes set up, it's time to connect them sequentially into a single, unified chain.

14. You are now ready to link this sequence to the **Attribute Noise** node, thereby completing the full PCG functionality.

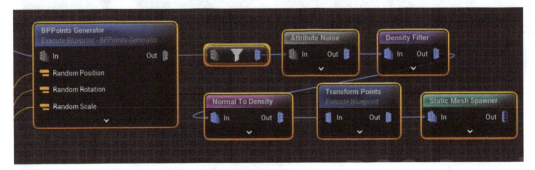

Figure 4.25 – A final look at the PCG graph

15. We have prepared our grass model at the beginning of our chapter. Let's use it by spawning grass around the rocks. Add the **Static Mesh Spawner** node inside the PCG graph.

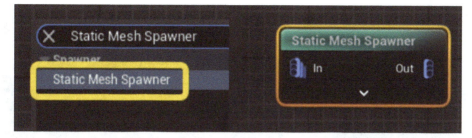

Figure 4.26 – Adding Static Mesh Spawner to the PCG graph

16. With the **Static Mesh Spawner** node selected, on the right-hand side, under **Mesh Entries**, press the + sign button to add one array element.

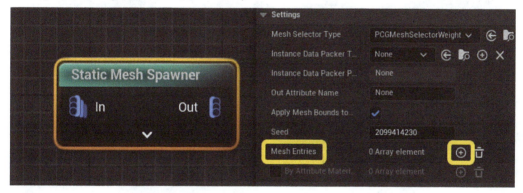

Figure 4.27 – Adding a Mesh Entries array

17. Add the SM_PlantGroup01 plant model to the **Static Mesh** entry array.

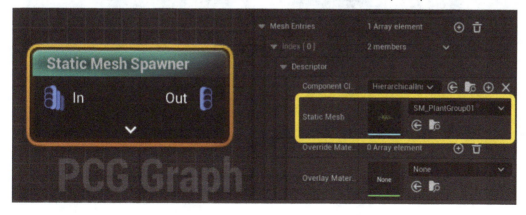

Figure 4.28 – Adding the SM_PlantGroup01 static mesh to the Mesh Entries array

18. Connect the **Static Mesh Spawner** node to the **BBPoints Generator** node so they are placed on top of each other, as is shown in the following figure:

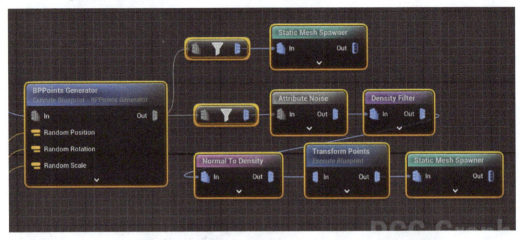

Figure 4.29 – A final look at the completed PCG graph

Now, we have reached the end of this exercise and you can check out the difference it makes after some modifications that were implemented into the PCG graph. Compile your PCG graph and let's check the results on the scene.

Figure 4.30 – View of the final results of BP_MeshGenerator on the scene

As observed, there are plants growing around each spawned rock. To enhance the appearance, you can adjust the values you've exposed in the Actor Blueprint. Additionally, utilize the Get Actor Property nodes within the PCG graph for further refinements. This combination allows for a more dynamic and interactive approach to changing the values via the editor scene.

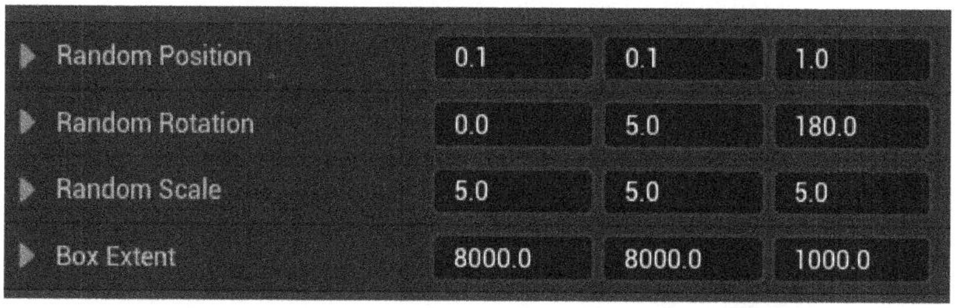

▶ Random Position	0.1	0.1	1.0
▶ Random Rotation	0.0	5.0	180.0
▶ Random Scale	5.0	5.0	5.0
▶ Box Extent	8000.0	8000.0	1000.0

Figure 4.31 – Value settings used for the current setup of BP_MeshGenerator

You've successfully completed this exercise. In the next section of this chapter, we'll explore optimization techniques and discuss how to enhance the efficiency of this PCG tool for your benefit. This will involve strategies to make the tool more effective and resource-efficient for your specific needs.

Optimizing and improving the PCG tool

This section is as important as the other mentioned in this chapter because it will help you understand the techniques and enable you to provide detailed information on any further optimization needed to improve the quality of the PCG tool's performance.

For this part, it will be great to go through each section of the PCG graph that procedurally generates static meshes within the PCG volume.

Foliage

This section is vital, as it's well known that foliage is often a major factor in reducing performance in game development and in VFX/virtual production setups. Therefore, it's essential to provide additional tips and techniques to enhance the efficiency of your PCG. By optimizing foliage and other procedural elements, you can ensure smoother gameplay without performance hiccups. Upcoming content will focus on these optimization strategies, helping to maintain high-quality visuals while ensuring the game or scene remains playable and efficient.

To begin with, let's explore a few settings that will help to visualize certain areas on the scene that require any further optimization:

1. In the top-left corner of the screen, you can find the **Lit** mode, which is on; press on it and choose **Virtual Shadow Map | Cached Page**.

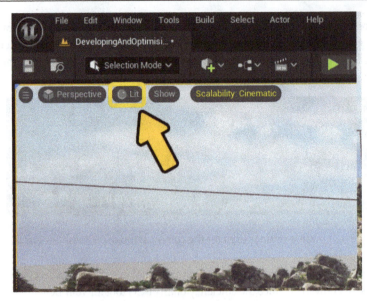

Figure 4.32 – View mode location

Figure 4.33 – Changing from Lit to the Cached Page mode

Figure 4.34 – The Cached Page mode view results

2. By checking those settings, you can now have full visuals on the **Virtual Shadow Maps** (**VSMs**) across this scene. Those in blue are our areas of interest because that's the main focus of our upcoming optimization process.

> **Note**
>
> **VSMs** in Unreal Engine 5 represent a significant advancement in shadow rendering technology. They are a part of Unreal Engine's **Lumen**, a fully dynamic global illumination and reflections system designed to achieve high levels of realism and visual fidelity in real-time rendering. That includes High-Resolution Shadows, Scalability, Dynamic Lighting, Integration with Lumen, Soft Shadows, and Performance Optimization.
>
> For more information, check out the documentation from the Unreal Engine 5 website:
>
> ```
> https://docs.unrealengine.com/5.4/en-US/virtual-shadow-maps-
> in-unreal-engine/
> ```

3. Foliage comes with the material nodes that are responsible for the movement of the plants. This is due to the **World Position Offset** (**WPO**) attribute that allows the translation of the vertices on mesh in runtime via a shader. This technique can be quite expensive in terms of performance, especially when you have lots of foliage on the surface. To reduce its complexity, let's open the PCG graph and select the **Static Mesh Spawner** node.

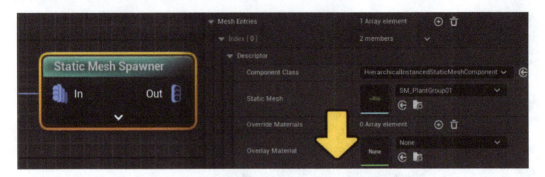

Figure 4.35 – Searching for the World Position Offset Disable Distance settings

4.   Under the **Mesh Entries** tab, navigate down until you find the section that says **World Position Offset Disable Distance**, as shown in *Figure 4.36*

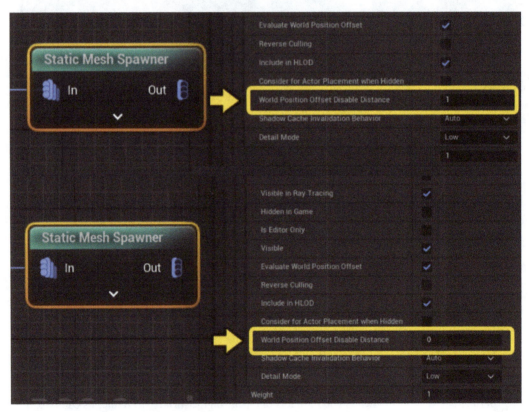

Figure 4.36 – Changing the World Position Offset Disable Distance value

5.   Currently, the setting is at 0, indicating a default distance. This means that the WPO is configured to render throughout the entire scene within the camera's reach. If you change the value to 1,

that means it will only render WPO within 1 Unreal Unit, which is equivalent to a distance of 1 cm. Let's give it a try and test that out!

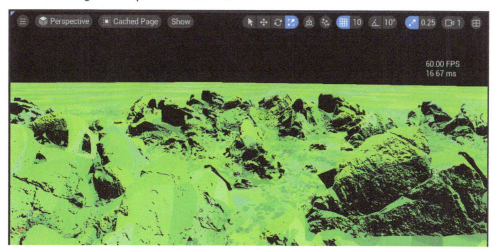

Figure 4.37 – Results after changing the World Position Offset Disable Distance value to 1

6. Let's head back to the main editor scene and find out what happened to our scene. As you can see now, everything turned out to be a green color because all the WPO has been reduced to only rendering in a distance of 1 cm.

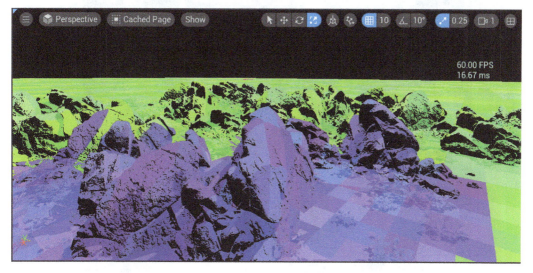

Figure 4.38 – Changing the World Position Offset Disable Distance value to 300

7. Let's try again, but this time, we will increase the distance by 300 centimeters. Now, you should see the WPO being rendered within those 300-centimeter distance ranges!

We just finished with the optimization for the foliage static meshes. In the next section, we will delve into additional techniques aimed at linking extra variables to the Actor Blueprint. These methods are designed to be efficient and effective, allowing for quicker and simpler adjustments without overburdening the PCG graph. By exploring these strategies, you'll gain insights into how to enhance your PCG setups in a way that is both performance-conscious and time-efficient.

Additional variables for the PCG tool

In this section, as the title suggests, we will get into the creation of additional variables inside an Actor Blueprint and PCG graph. We will focus on adding a few key variables that are instrumental in enhancing the external functionality of your PCG tool. By incorporating these variables, you'll be able to extend the capabilities of your PCG tool, allowing for greater flexibility and control in your procedural content creation. This approach is designed to make your PCG tool more dynamic and adaptable to various scenarios and requirements:

1. Let's open your `BP_MeshGenerator` Actor Blueprint, add a Boolean variable, and rename it `Absolute Rotation`. Do the same but add vector variables and rename them `OffsetMin` and `OffsetMax`. Always remember to make them visible for each variable. Compile and let's head back to the PCG graph.

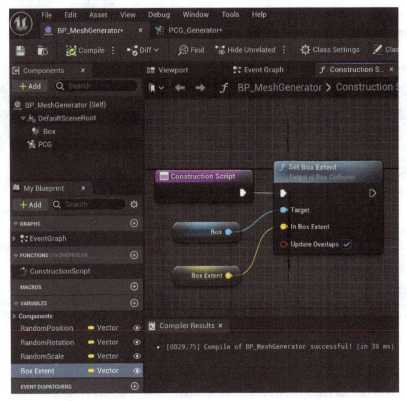

Figure 4.39 – The VARIABLES panel window

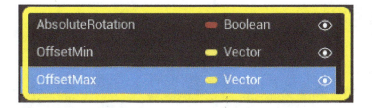

Figure 4.40 – Adding variables to the VARIABLES panel window

2. Inside the PCG graph, you need to expose those variables. So, let's add the **Get Actor Property** nodes to your PCG graph.

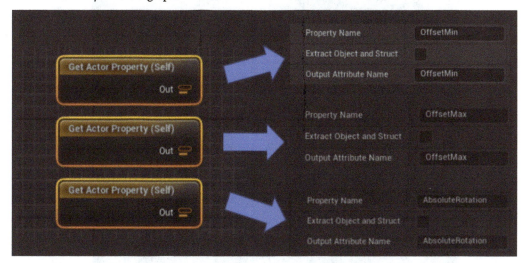

Figure 4.41 – Renaming the Get Actor Property parameters

3. Each **Get Actor Property** node is ready to become a variable and now you can connect to the **Transform Points** node, which you will use to manipulate the transform for each spawned grass on top of the rock. Click on the arrow to expand the **Transform Points** node.

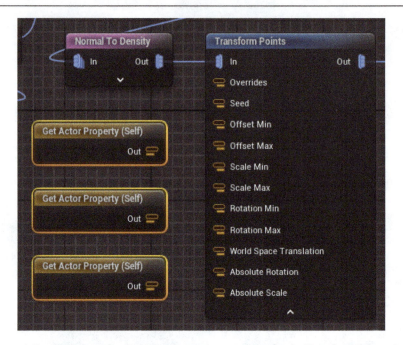

Figure 4.42 – Setting up the Get Actor Property parameters for the Transform Points node

Here is the expanded **Transform Points** node, where we can accurately connect our **Get Actor Property** nodes to the appropriate input properties.

4.   Connect each **Get Actor Property** node accordingly so it matches the names of each variable.

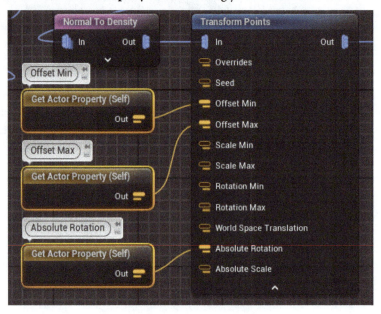

Figure 4.43 – Connecting the Get Actor Property parameters

With this part completed, save your PCG graph, and let's test it on your scene. This way, you will be able to tweak the values in the editor scene without interrupting the PCG graph. As you may have noticed, those are the three variables that you exposed in your Actor Blueprint and connected inside your PCG graph.

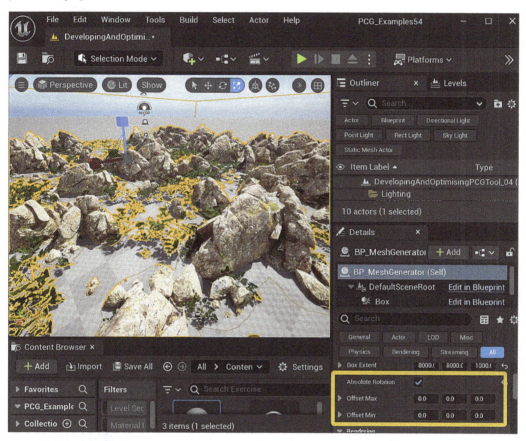

Figure 4.44 – The final adjustments on the BP_MeshGenerator details panel inside the level editor

We have now reached the end of this chapter. You have gained valuable knowledge on developing a highly useful PCG tool that includes additional optimization and properties for fine-tuning values, thereby enhancing the overall quality of your PCG tool. This is a crucial step in preparing you to better understand, interact with, and fix the PCG at your convenience.

Summary

In this chapter, you've gained valuable insights into the significant benefits of enhancing your custom PCG tool. You've learned how to utilize and optimize this tool to align with your workflow, particularly in the context of project development. In this case, you have learned how to improve the integration of the PCG Blueprint into the PCG graph and you added a bunch of variables that help to improve the workflow of your PCG tool! You also covered the optimization tools available within the PCG graph and its nodes that help to improve the speed of rendering the spawned static meshes on the landscape. This knowledge is crucial as it empowers you to further improve your PCG tools, especially when dealing with the creation of more complex procedurally generated content.

After finishing this tutorial on creating and fine-tuning your PCG tool, it's advisable to review both the previous chapters and the current one again. This time, try to apply what you've learned using your own models and configurations. The primary objective of this chapter is to equip you with the knowledge and skills needed to enhance your PCG tool's performance. By the conclusion of this chapter, you will have gained practical experience in crafting your own PCG tool, which includes designing a basic custom node and incorporating it into the PCG graph with different nodes and settings.

In the upcoming chapter, we will shift our focus to the creation and utilization of a PCG tool that uses a spline controller. We'll explore the key advantages of using a spline-based approach in PCG tools. This method offers enhanced control and precision in the placement and manipulation of procedural content, making it an invaluable technique for certain types of projects and design goals. I cannot wait to explore this next chapter together!

Part 2:
Harnessing the Power of PCG
in Unreal Engine 5

In this section, you will get into the creation of dynamic PCG environments using spline controllers, along with advanced techniques for crafting diverse landscapes, including the integration of structures, all through the use of spline components. You'll develop a deep understanding of PCG and explore how to leverage components within actor blueprints to enhance a PCG framework. As we progress, we'll tackle even more exciting examples, such as crowd simulations, to further expand your skill set.

This part has the following chapters:

- *Chapter 5, Building Spline Controllers with PCG Graph*

- *Chapter 6, Building a PCG Graph with Landscape Materials*

- *Chapter 7, Let's Build a Building Using the PCG Spline Controller*

- *Chapter 8, Building Biomes: Mastering PCG for Rich Environments*

- *Chapter 9, Creating Dynamic Animated Crowds with PCG*

5

Building Spline Controllers with PCG Graph

In *Chapters 3* and *4*, you made remarkable progress by mastering the utilization of your own **Procedural Content Generation (PCG)** blueprint. This technique empowers you to customize PCG components, thereby constructing your personalized PCG tool, including the techniques that aid in optimizing your static mesh assets

Chapter 5 presents a distinct approach to PCG, offering an engaging opportunity to delve deeper into understanding a PCG tool. This time, we will employ spline controllers to achieve outcomes akin to those in previous chapters, such as *Chapters 2* and *3*. Utilizing spline controllers is pivotal in this book, providing comprehensive and flexible control over your PCG tool.

Within this chapter, you'll get into crafting a small forest with a river body flow. Additionally, you'll integrate the Unreal Engine water plugin to create rivers and discover how to manipulate them within Unreal Engine 5.

This chapter is divided into sections focusing on various PCG graphs, which we will subsequently combine into a single PCG graph. Following this, we will complete the remaining configuration within this unified PCG graph before integrating it into a single Actor Blueprint.

You'll gain proficiency in selecting the appropriate tools for worldbuilding within Unreal Engine, a skill that enhances your ability to construct more intricate worlds and levels directly within the scene.

This chapter takes a deeper look at the following topics:

- Creating foliage
- Creating the forest Actor Blueprint
- PCG graphs
- Making adjustments for better outcomes

After completing this chapter, you will gain new techniques and very valuable knowledge on creating a PCG tool that can be controlled using the spline controller! Upon finishing this hands-on tutorial, you will have the ability to control and customize the shape of your PCG Volume and procedurally generate any models on the landscape

Technical requirements

You will need the following hardware and software, as well as knowledge, to complete this chapter:

- A computer that can run a PCG with the following specifications is essential:

  - A high-performance multi-core CPU (AMD Ryzen 7/9, Intel i7/i9)

  - GPU (NVIDIA RTX, AMD Radeon RX with 8+GB VRAM)

  - At least 16 GB of RAM

  You must have basic knowledge of blueprint creation (`https://dev.epicgames.com/documentation/en-us/unreal-engine/blueprint-basic-user-guide-in-unreal-engine`) in Unreal Engine 5. This is crucial, especially for those who are new to blueprint creation. The chapter is designed to be accessible to those with even a fundamental understanding of blueprint development.

  The template project comes without the assets apart from the house cabin. Make sure to download *Megascans Trees: European Beech* and the *Landscape Pro 2.0 Auto-Generated Material* project files from the Unreal Marketplace (`https://www.unrealengine.com/marketplace/en-US/product/landscape-pro-auto-generated-material?sessionInvalidated=true`). All of those assets are free of charge.

- The Unreal Engine version that's used in this chapter is version UE 5.4. The PCG tool was introduced with Unreal Engine 5.2 but there are some nodes that were deprecated in Unreal Engine 5.4. Hence, the latest version of Unreal Engine is the most beneficial one for this chapter.

For this chapter, we will use the template project, which is available in the GitHub repository. You can download its template at `https://github.com/PacktPublishing/Procedural-Content-Generation-with-Unreal-Engine-5`.

> **Note**
> GitHub repository doesn't include any models from the marketplace and you will need to download the assets separately into the project from GitHub.

The code in action video for the chapter can be found at `https://packt.link/H1Bxh`

Creating foliage and the forest Actor Blueprint

In the following section, we will guide you through preparing the essential assets for this chapter. We'll start with creating foliage, including selecting and customizing plant and tree models. Next, we'll move on to setting up the forest actor blueprint, which defines the behavior and appearance of the forest elements. By the end of this section, you'll have a complete set of assets and a functional blueprint to start work with the PCG forest environment.

Creating foliage

For this exercise, we will *Megascans Trees: European Beech* package from the marketplace. However, this time, we will diversify the trees a little bit to build your PCG graph actor blueprint. We will use the standard grass that you have already used in the previous chapter; it comes from *Landscape Pro 2.0 Auto-Generated Material* from the marketplace. Let's pick up a few models from our existing library of trees inside the Content Folder:

1. Navigate to **EuropeanBeech | Geometry | PivotPainter** and search for the two tree models that are called SM_EuropeanBeech_Field_01_PP and SM_EuropeanBeech_Forest_03_PP.

Figure 5.1 – Quixel Megascan tree models

2.   Navigate to **STF** | **Pack03-LandscapePro** | **Environment** | **Foliage** | **GrassBush** and search for SM_grass_bush04_lod00.

Figure 5.2 – Pack03-LandscapePro plant asset

3.   The next step of this part is to turn on **Nanite** mode for each mesh. With the models selected, right-click on one of the models and check **Nanite** mode at the top of the menu.

Figure 5.3 Enabling Nanite mode for each static mesh

In the next section, we'll dive into creating the actor blueprint, where we'll prepare an actor blueprint for the PCG forest foliage.

Creating the forest Actor Blueprint

In this part of the chapter's tutorial, we will be taking time to create an essential actor blueprint. This will be used to control and procedurally generate the forest based on the PCG component, much like what you have already done in *Chapters 3* and *4*.

To begin with, let's create your first Actor blueprint inside the project and prepare the components that you are going to use to control later within the PCG graph:

1. Inside the **Blueprints** folder, right-click on the content browser space and search for **Blueprint Class**.

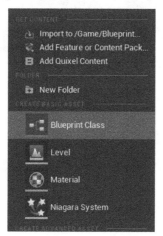

Figure 5.4 Selecting Blueprint Class

2. With **Blueprint Class** selected, inside the menu, choose the **Actor** class and rename it to BP_Forest.

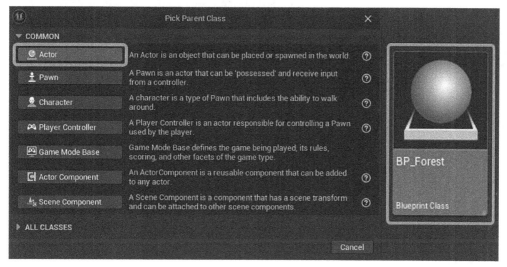

Figure 5.5 Renaming the Actor blueprint to BP_Forest

3. Now double-click on your newly created BP_Forest actor blueprint and then start adding the **Spline** and **PCG** components to it from the +**Add** menu.

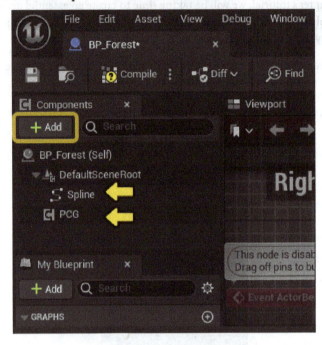

Figure 5.6 – Adding the Spline and PCG components to the BP_Forest Actor blueprint

4. The following step is critical as it will be used to control and manage the components of the PCG graph that influence the quantity of foliage spawned within the PCG Volume. Therefore, let's create two float variables named **Trees** and **Grass**. Make sure that both of these variables are set to be public. You can turn on the visible eye icon on the right-hand side of each variable

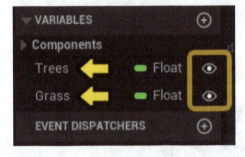

Figure 5.7 – Creating and preparing the variables

5. With these variables, you are all set. However, we still have to do one more thing, and that is to change the values of each created variable. For each variable, let's set the value to 1000. The purpose of using these values is to eliminate any performance issues while testing the Actor Blueprint.

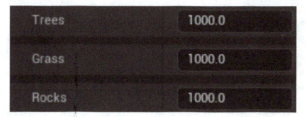

Figure 5.8 – Tweaking the values for each variables

6. Let's select the spline component and create the spline curve loop, similar to *Figure 5.9*. To do that, select the little pin at the end of this curve. While holding the left *Alt* button, use the left mouse button to start adding more points to the curve along the spline.

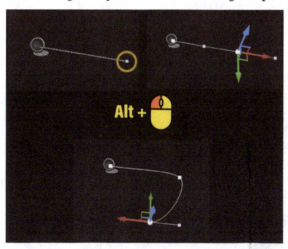

Figure 5.9 – Creating the spline loop using the spline component

7. Keep adding the spline points to your curve until you create a circular loop shape.

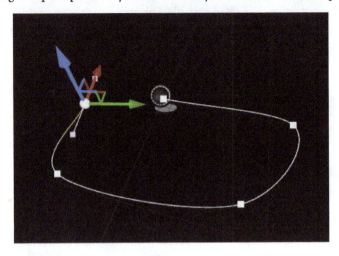

Figure 5.10 – Adding more spline points around the spline loop

8. Once your spline curve is ready, check the **Closed Loop** box on the **Spline** details panel to close the loop.

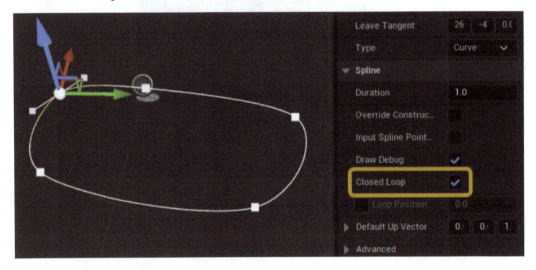

Figure 5.11 – Closing the loop for the spline component

9. Now we can close our `BP_Forest` actor blueprint. Let's take this chance to drag and drop our blueprint into the scene.

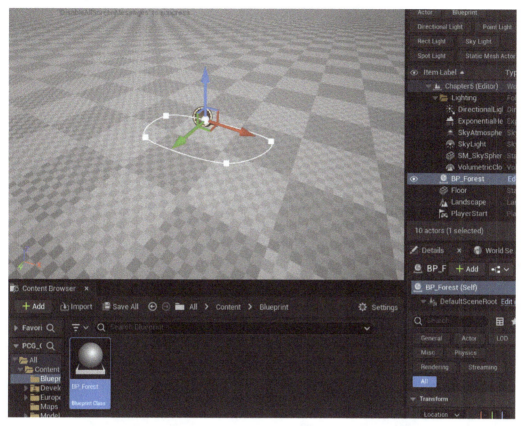

Figure 5.12 – Adding the BP_Forest Actor Blueprint to the level scene

In the upcoming section, we'll explore how to create several PCG graphs and the methods for integrating them with one another.

Creating PCG graphs

In this exercise, you will develop several PCG graphs and link them to create a network chain culminating in a single main PCG graph. This activity is designed to highlight the importance of combining multiple PCG graphs in order to make it look more clean and less messy.

To begin our exercise, let's use the PCG folder and add a few PCG graphs inside there.

Right-click on the content PCG folder space and create a new PCG graph. Duplicate the PCG graph and name the graphs individually: PCG_Foliage and PCG_Forest.

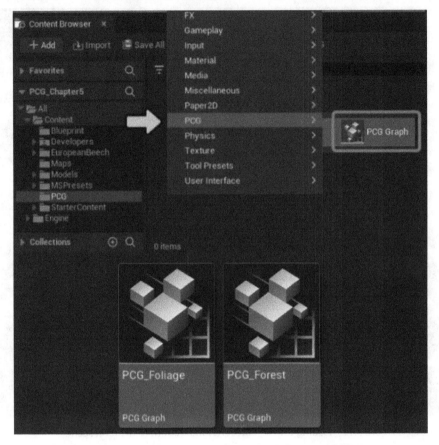

Figure 5.13 – Creating two PCG graphs – PCG_Foliage and PCG_Forest

We will go one by one, working on and testing each individually until both of them are ready to be combined into one PCG graph. Let's get started with the first one: PCG_Foliage!

PCG_Foliage

In this section, we will create PCG nodes for PCG_Foliage and develop the logic for spawning foliage based on its position and encountered obstacles throughout the landscape. The following steps include adding spline nodes and nodes to calculate the distances between the spawned static meshes and the obstacles they face along the terrain:

1. Let's open the PCG_Foliage graph and start adding the main components to the graph space. Right-click and search for Get Actor Data.

Figure 5.14 – Adding Get Actor Data inside the PCG graph

2. Next, we will be adding two **Spline Sampler** nodes to the graph. Each of them will be composed of different settings to improve the quality of the procedurally generated static meshes onto the scene.

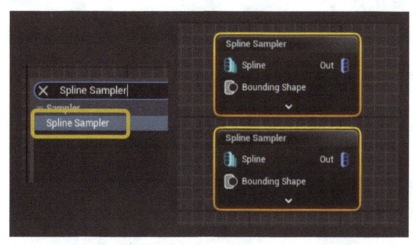

Figure 5.15 – Adding two Spline Sampler nodes in the PCG graph

3. Choose the topmost **Spline Sampler** node. Let's make several adjustments within its **Settings** panel. For this sampler, set **Dimension** to **On Interior**. This modification ensures that static meshes are spawned within the closed loop instead of outside. Remember to check the **Unbounded** option; failing to enable this feature will prevent the static meshes from spawning correctly.

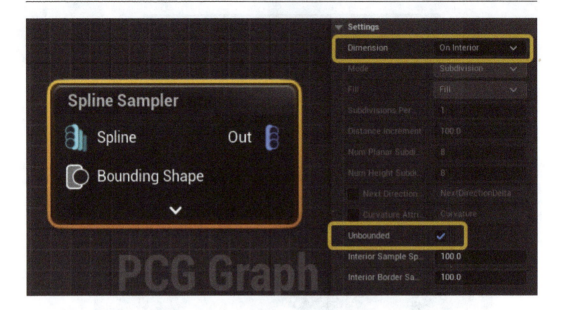

Figure 5.16 Enabling the Dimension and the Unbounded settings for the Spline Sampler node

4. Now, select the other **Spline Sampler** node and modify its settings. This time, it will behave
    in contrast to the first one, which means that the static meshes will be spawned on the outer
    layers of the spline curve. Change **Dimension** to **On Spline** and make sure the **Unbounded**
    box is checked!

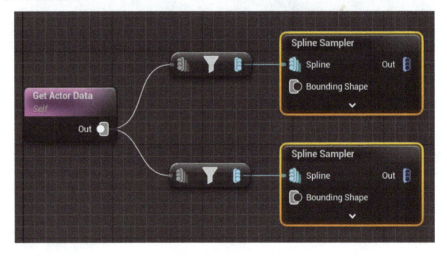

Figure 5.17 – Connecting two Spline Sampler nodes to the Get Actor Data node

5.   Let's connect both **Spline Sampler** nodes together with the **Get Actor Data** node.

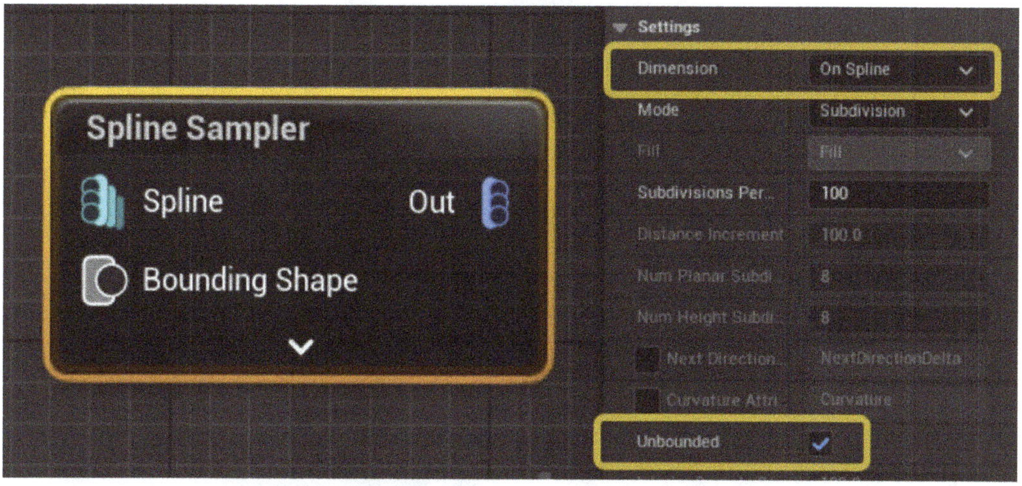

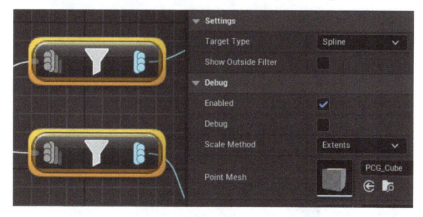

Figure 5.18 – Setting up the second Spline Sampler with the On Spline setting

6.   Select both **Filter – Spline** nodes that are linked between the main nodes and adjust their **Target Type** in the right **Settings** panel from **Curve** to **Spline** mode. This modification enables the **Get Actor Data** node to function in tandem with the **Spline Sampler** nodes.

Figure 5.19 – Changing the Target Type setting to the Spline type

7.   Let's add and introduce a new element to the PCG graph in this chapter: the **Distance** node. This node plays a unique role in PCG mechanics. It is tasked with calculating and attaching a distance attribute to each source point. This calculation is performed between the source point and its closest target point.

Figure 5.20 – Distance node components

8. With the **Distance** node set up, we don't have to make any changes inside the **Settings** panel. Let's connect the **Spline Sampler** node accordingly where the top node connects with the **Source** input and the bottom node connects with the **Target** input.

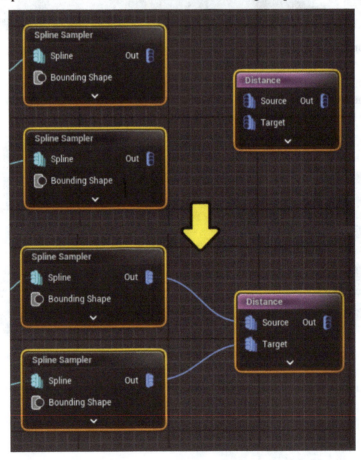

Figure 5.21 – Connecting two Spline Sampler nodes to the Distance node accordingly

9. At this point, you can continue with the graph. Let's add the **Transform Points** and connect it to the **Distance** node!

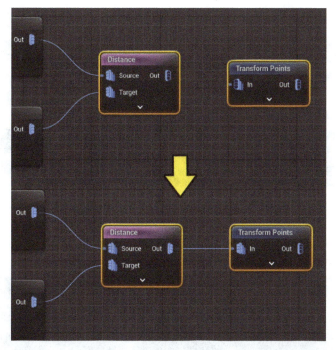

Figure 5.22 – Connecting the Distance node to the Transform Points node

10. The following step involves repeating *Steps 1-10*, but with a key difference: we will utilize just a single **Spline Sample** node. Let's add the **Get Actor Data** node to our graph again. Let's also add a **Spline Sampler** node and connect them.

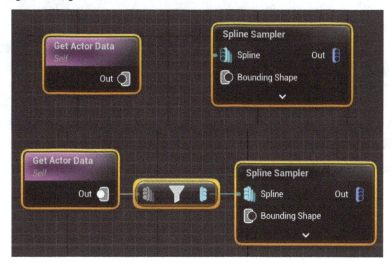

Figure 5.23 – Connecting the Get Actor Data node to the Spline Sampler node

11. Before we can move on to adding more nodes, we need to set a few settings. For this, let's go and change the **Filter** inner target type to the **Spline** type.

Figure 5.24 – Changing the target type from Curve to Spline

12. Select the **Spline Sampler** node and change its **Dimension** setting to **On Interior**. Check the **Unbounded** box.

Figure 5.25 – Enabling the Unbounded setting in the Spline Sampler settings menu

13. With the nodes set, let's add **Transform Points** to its mix and connect it together with the **Spline Sampler** node.

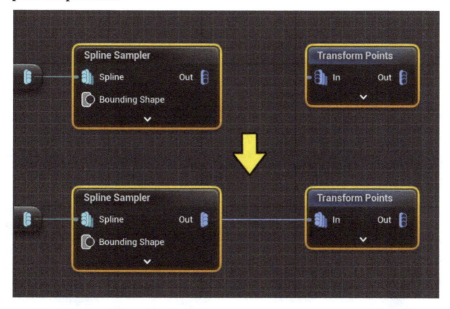

Figure 5.26 – Connecting the Spline Sampler node with the Transform Points node

14. The next step is very crucial and will involve connecting these nodes with the **Output** node. However, for this to happen, we need to configure and add the output pins inside the **Output** node. Search for the **Output** node on your PCG graph; it should be somewhere near the top. Then drag it down and place it next to your newly created node structure.

Figure 5.27 – Moving the Output node closer to the main graph

15. After dragging the **Output** node closer to the main graph with all the nodes, we can start customizing the **Output** node with new pins.

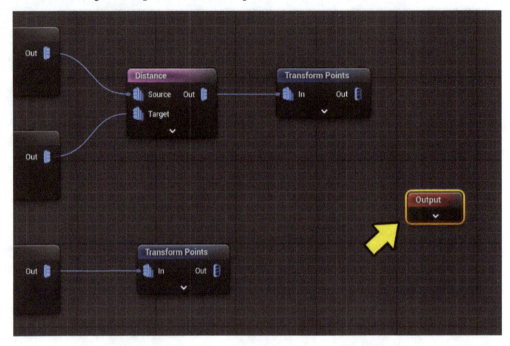

Figure 5.28 Placing the Output node next to the existing nodes

16. The **Output** node offers good flexibility, allowing you to add additional output pins tailored to the specific functions of the pins. In this instance, we'll utilize this node to link the **Output** type, ensuring it aligns with the points that are generated inside the other PCG graphs. For now, let's add two custom pins in the **Input** settings.

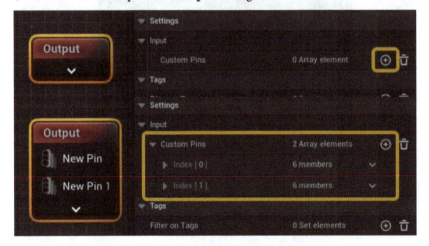

Figure 5.29 – Adding two pins inside the Output node

17. Open each new custom pin and rename each pin to `Trees` and `Grass`.

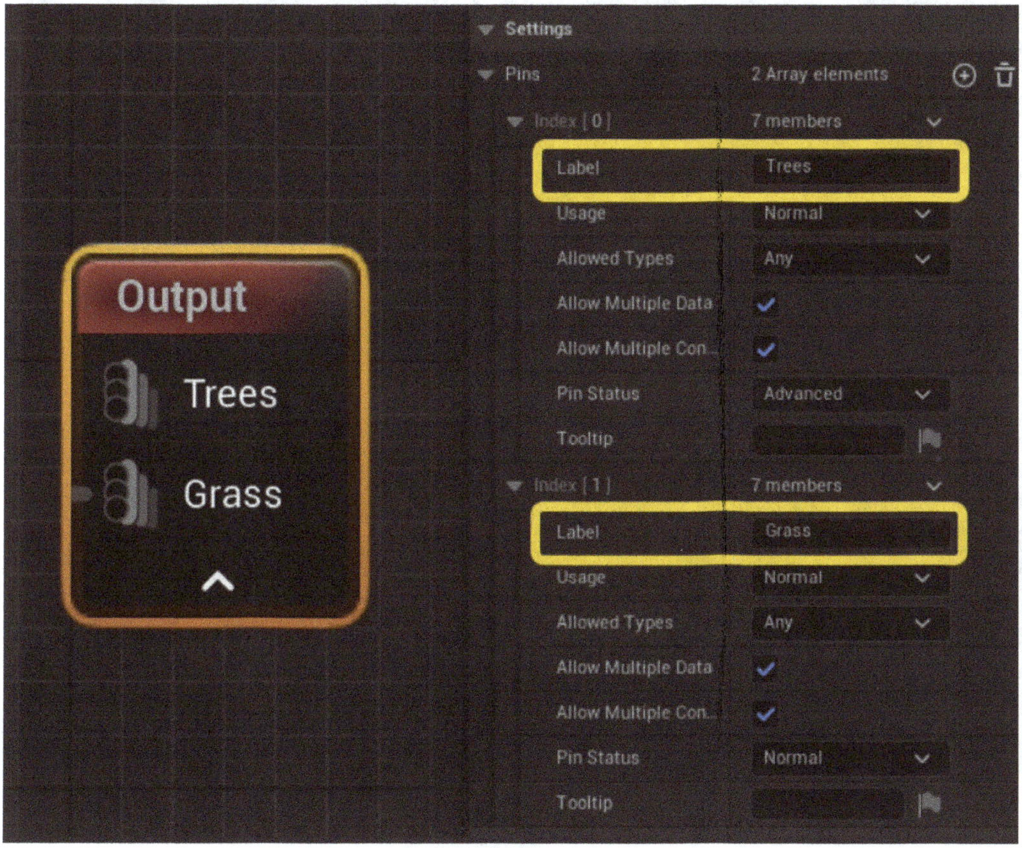

Figure 5.30 – Changing the Label names for the Output pins

18. Set **Allowed Types** from **Any** to **Point** for each one. You should see the pins changing their colors to blue inside the **Output** node!

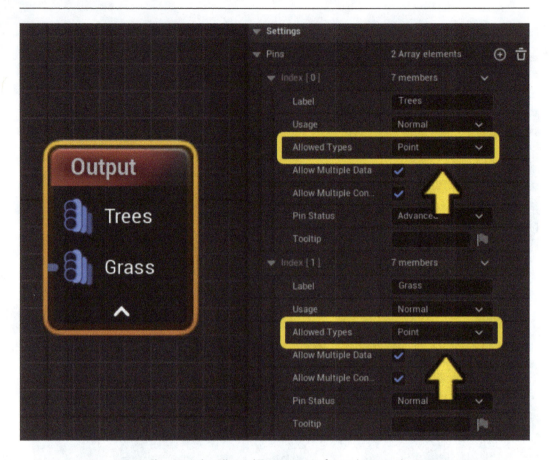

Figure 5.31 – Changing the Allowed Types setting for each pin to the Point type

19. At this stage, we have added all the nodes and we need to finish with the foliage PCG graph. Now we can connect each graph with the **Output** node.

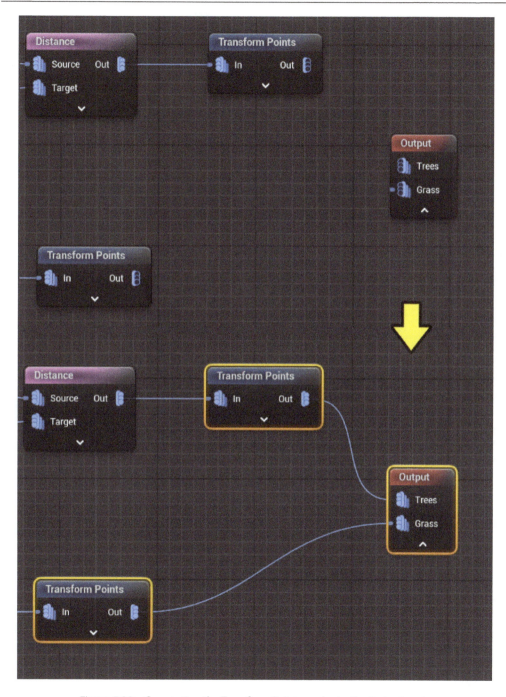

Figure 5.32 – Connecting the Transform Points nodes to the Output node

20. We're nearing the completion of the foliage PCG graph, but a few additional nodes are still left to end this section. To accomplish this, place a **Get Actor Property** node above each **Spline Sampler** node in your PCG graph, following the arrangement shown in the *Figure 5.33*. The purpose of using these nodes is to provide better control over the PCG graph via the Actor blueprint on the scene.

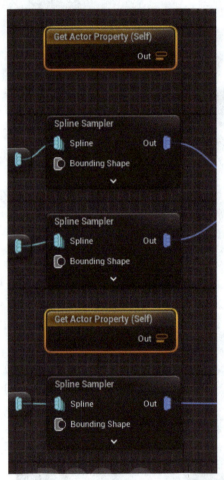

Figure 5.33 – Adding the Get Actor Property nodes next to the Spline Sampler nodes

21. The next step requires adjusting the property name for each node graph to align with Trees and Grass, respectively. Each **Spline Sampler node**, designated as 1 for Trees and 2 for Grass, should be configured to reflect the variables defined in the Actor blueprint at the chapter's outset.

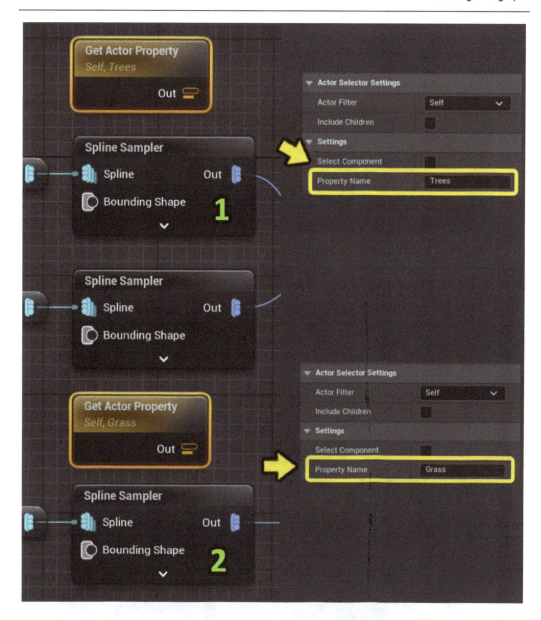

Figure 5.34 – Changing the Property Name value for the Trees and Grass variables

22. Now, you should link the **Get Actor Property** nodes to the appropriate property function located within the **Spline Sampler** node. The specific function we're targeting is named **Interior Sample Spacing**. This function ensures that points are evenly distributed, providing space for spawning static meshes in between.

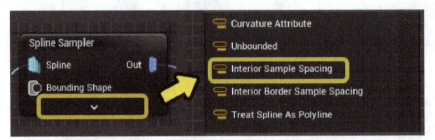

Figure 5.35 – Expanding the Spline Sampler node to find the Interior Sample Spacing input

23. With that node expanded, you can connect the **Get Actor Property** node to this function.

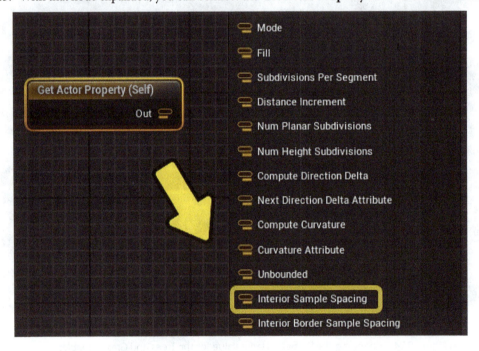

Figure 5.36 – Connecting Get Actor Property to the Interior Sample Spacing input

24. You can minimize and close your node by pressing the arrow that points upward at the bottom of the **Spline Sample** node.

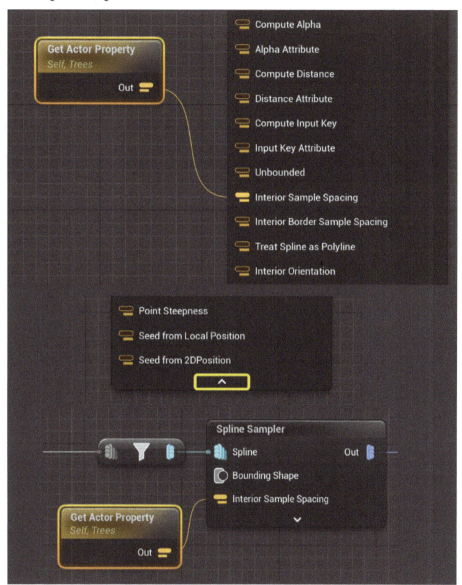

Figure 5.37 – Minimizing the Spline Sampler node and reducing its size in the graph

25. Let's repeat the same process with another **Spline Sampler** node. This time, connect the **Get Actor Property** node for Grass functionality.

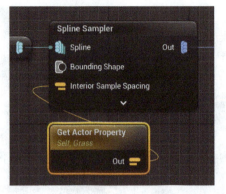

Figure 5.38 – Spline Sampler node together with the Get Actor Property node

That's how your graph should look after connecting all the nodes to the right positions (*Figure 5.39*), with 1 for Trees and 2 for Grass.

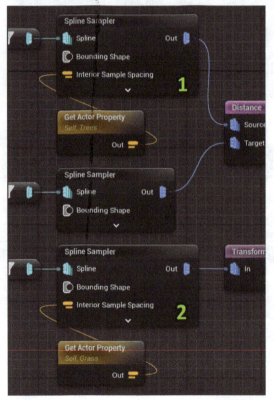

Figure 5.39 – Two Spline Sampler nodes for controlling the amount of grass and trees

26. Here is the final look at the PCG graph. For this, we placed all the nodes very close together for a better view (*Figure 5.40*).

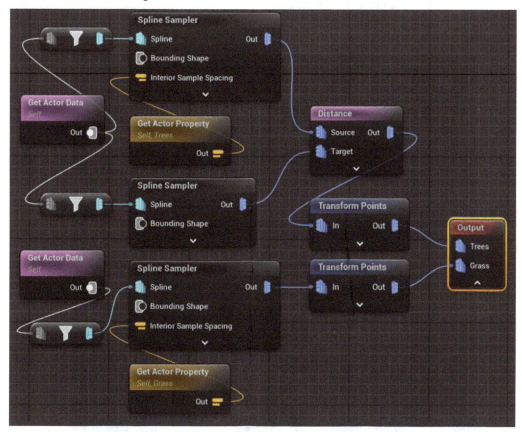

Figure 5.40 – A visual look at the logic applied in the section

You've just completed the section on creating the foliage PCG graph, where you were introduced to the fundamentals of developing spline sampler logic. While there are still a few PCG graphs left to finalize the whole tool, we're now at a point where we can begin integrating it into another PCG graph, specifically PCG_Forest. In this next part, we'll enter the testing phase for PCG_Foliage by utilizing the PCG_Forest graph.

Testing PCG_Foliage

Next, we'll start the process by opening the PCG_Forest graph and merging it with the PCG_Foliage graph. This will serve as a fascinating test to examine how well different PCG graphs can be integrated.

Let's start by locating your `PCG_Forest` graph:

1. This is the right time to try and test the foliage PCG graph configuration. For this, we will use a `PCG_Forest` graph and integrate a `PCG_Foliage` graph. Let's open the `PCG_Forest` graph inside the `PCG` folder.

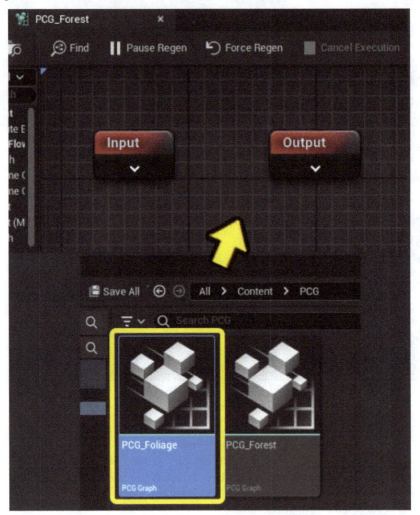

Figure 5.41 – Drag and drop PCG_Foliage to the PCG_Forest graph

2. With the `PCG_Forest` graph open, let's drag and drop the `PCG_Foliage` graph into the `PCG_Forest` graph space; see *Figure 5.42*.

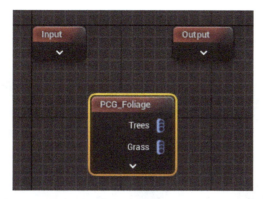

Figure 5.42 – PCG_Foliage subgraph inside the PCG_Forest graph

3.   At this point, we are going to test the functionality of the `PCG_Foliage` graph. To do that, let's add two **Static Mesh Spawner** nodes and place them next to the **PCG_Foliage** node. For now, let's connect each node with the corresponding name.

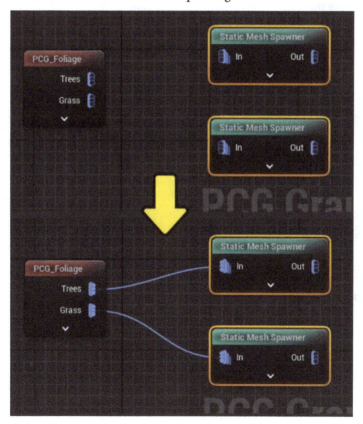

Figure 5.43 – Connecting two Static Mesh Spawner nodes to the PCG_Foliage node

4. At the moment, there is not much going on because we haven't assigned any static mesh to the corresponding nodes yet. Let's start with the first top **Static Mesh Spawner** node and add two **Mesh Entries** arrays.

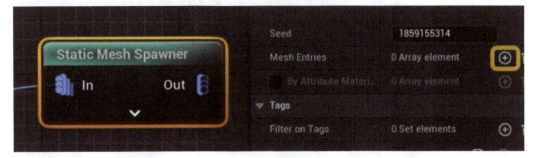

Figure 5.44 – Adding Mesh Entries in the Static Mesh Spawner node

5. For the **Static Mesh Spawner** node, we will choose the two **Trees** models that we prepped at the beginning of this chapter. Those are SM_EuropeanBeech_Field_01_PP and SM_EuropeanBeech_Forest_03_PP.

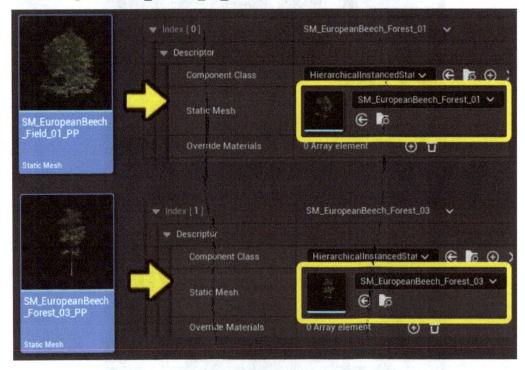

Figure 5.45 – Adding two tree static meshes to each mesh entry

6. Let's repeat the same process on the other **Static Mesh Spawner** node. However, this time we will use the Grass static mesh.

Figure 5.46 – Adding Mesh Entries in the Static Mesh Spawner node

7. We will add the standard grass that you have used in every chapter of this book, which is SM_grass_bush04_lod00.

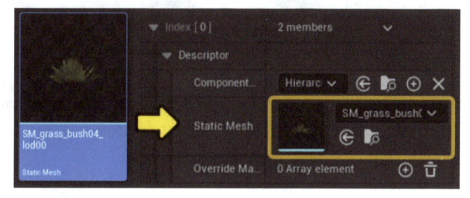

Figure 5.47 – Adding plant static mesh to the mesh entry inside the Static Mesh Spawner node

We just finished implementing the **Static Mesh Spawner** nodes inside the PCG_Forest graph. Now, it's time to move on to the next phase of testing, which is opening the BP_Forest actor blueprint.

With this in mind, let's head back to your BP_Forest actor blueprint and add PCG_Forest into the PCG component.

8.   Open your BP_Forest actor blueprint inside your content browser.

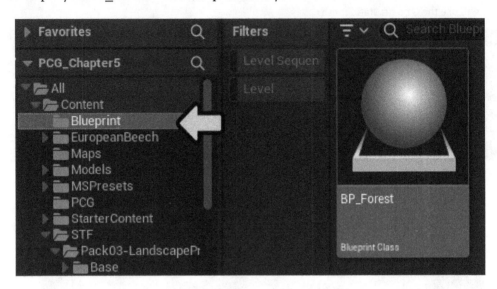

Figure 5.48 – Locating the BP_Forest inside the Blueprint folder

9.   Inside your actor blueprint, under the **Components** panel, select the PCG component and add the PCG_Forest graph to the **PCG Graph** slot on the right-hand side under the **Details** panel.

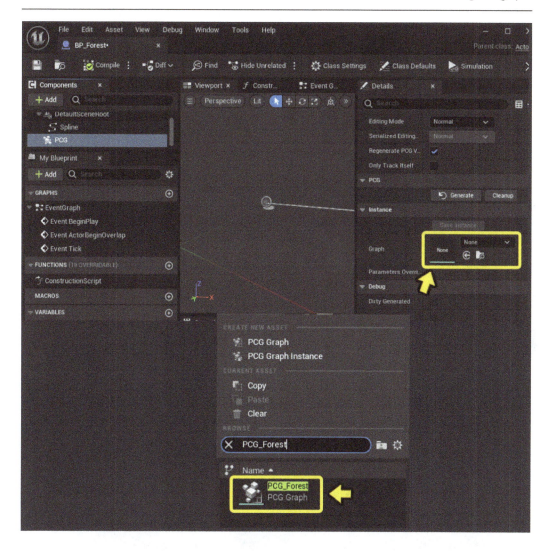

Figure 5.49 – Selecting the PCG component and adding the PCG graph on the Details panel

10. Once this part is done, compile your actor blueprint. Then, let's head back to your viewport and test out your PCG spline tool. Use **Spline Points** to extend the spline curve along the landscape surface.

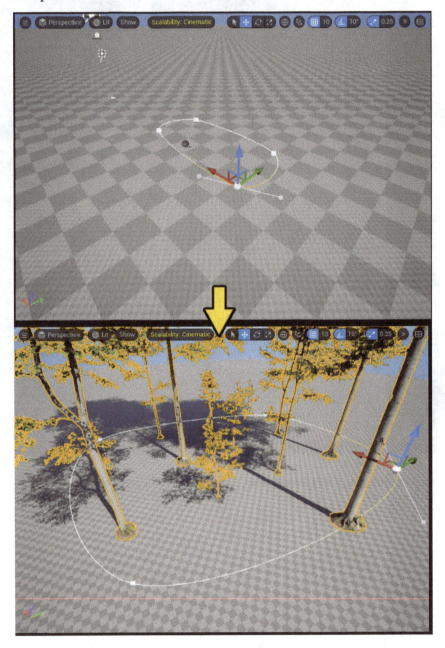

Figure 5.50 – Testing the BP_Forest spline Actor Blueprint

11. The current setup is looking better, but there's room to improve the amount of spawned foliage. Select the `BP_Forest` actor blueprint in the outliner, and then, on the right-hand side, adjust the settings inside the **Details** panel. We should modify the **Grass** values to around `100` while leaving the number of trees as it is.

Figure 5.51 – Reducing the Grass value to distribute the spawned grass static meshes closer together

The results are beginning to form and it adds more vibrancy to the scene. However, it's evident that some adjustments to the transformations are needed, as the grass appears too uniform in its orientation. We'll address this issue later in the chapter as we explore other nodes, ensuring a more natural and varied appearance.

Figure 5.52 – Irregular grass formation inside the BP_Forest Actor blueprint

In the next part of the chapter, we will guide you through the steps to add the **Water** plugin to the project. We will use one of the water body examples, specifically the river body. Then, I will demonstrate how to integrate the water body into the PCG graph so that the river flows through the forest.

PCG water

This part will be very similar to the previous exercise you did with the use of the **Get Spline Data** node. However, this time we will need to enable the plugin first and then refine the tag value to make sure it works within the PCG_Graph editor:

1. Let's open the **Plugin** tab inside Unreal Engine. Navigate to **Edit | Plugins**. In the **Plugins** window, search for Water. Enable all the Water plugins. It will ask you to restart now, so press the **Restart** button to restart Unreal Engine.

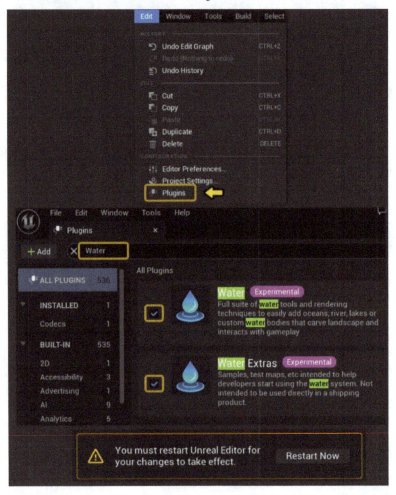

Figure 5.53 – Searching for the Plugins in the Edit menu and enabling a Water plugin

2. To make sure that the **Water** plugin works, let's test the water assets from the plugin. For this tutorial, we will use the **Water Body River** asset. In the main **Viewport** menu, click the icon with the + sign and click on the **Place Actors Panel** tab.

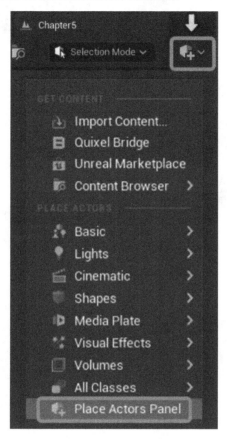

Figure 5.54 – Accessing Place Actors Panel

3. On the left-hand side, inside **Place Actors Panel**, enter Water Body River. Select the Water Body River and then drag and drop directly into the scene. You can test and modify it to match your preferences. We will use this asset later for this exercise!

Figure 5.55 – Searching for the Water Body River actor

4. Inside your PCG folder, let's add the PCG graph and call it PCG_Water. Open it so we can start adding the nodes we need for this exercise.

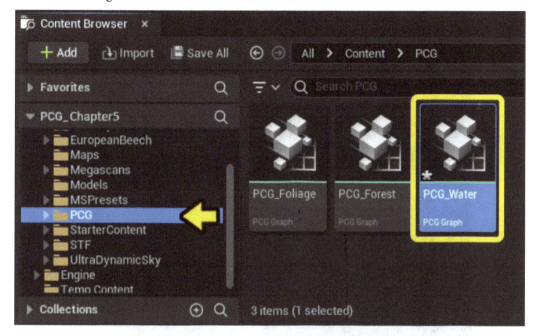

Figure 5.56 – Adding a new PCG graph and renaming it to PCG_Water

5. We only need two nodes for this exercise; those are **Get Spline Data** and the **Spline Sampler** node. Stack them next to each other.

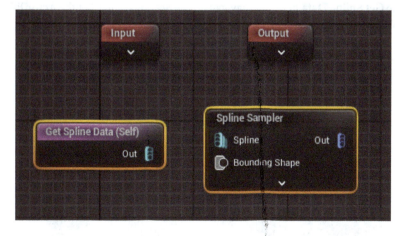

Figure 5.57 – Adding the Get Spline Data and Spline Sampler nodes

6.   Select the **Spline Sampler** node. On the **Settings** panel, change **Mode** from **Subdivision** to **Distance**. Turn on the **Unbounded** attribute.

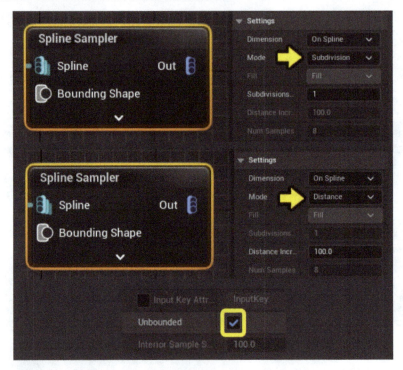

Figure 5.58 – Changing the Spline Sampler's mode to Subdivision and enabling Unbounded

7.   Select the **Get Spline Data** node and navigate to the **Settings** panel. Set the **Actor Filter** option to **All World Actors**.

Figure 5.59 – Changing the Actor Filter to All World Actors

8. The next part is to use the **Actor Selection**. However, we are not going to use **By Tag**. Instead, we will choose **By Class**. Since the water plugin is a built-in plugin, the only way to access it is via the existing class. In the **Actor Selection Class** tab, choose **Water Body River**.

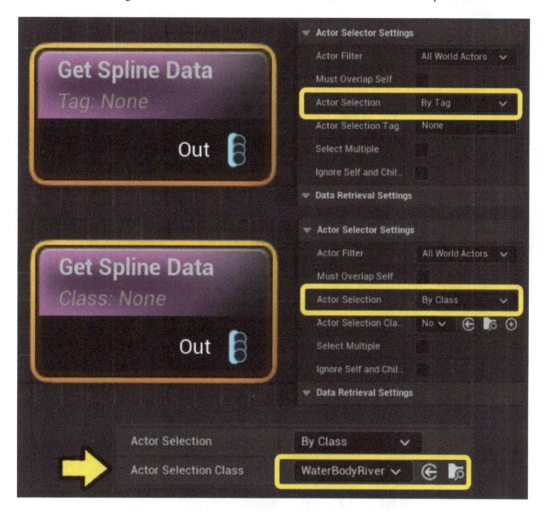

Figure 5.60 – Changing the Actor Selection to By Class

9. To finalize the water PCG setup, let's connect nodes, including the **Output** node at the end.

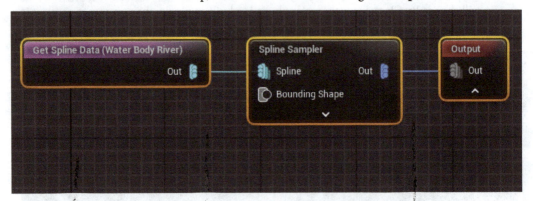

Figure 5.61 – Connecting all the nodes together

10. Before we can move to another section, we have to set the **Output** node to work with the water plugin. To do this part, select the **Output** node. On the right-hand side, under the **Settings** panel where the **Input** tab is, click on the + button to add the **Index** array element.

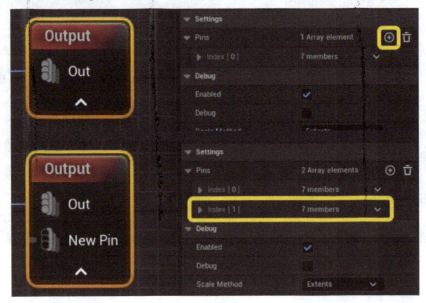

Figure 5.62 – Adding an Index array to the Output node

11. On the new pin index array, change your name to Water and change **Allowed Types** from **Any** to **Point**. You will notice that the pin has changed its color from grey to light blue.

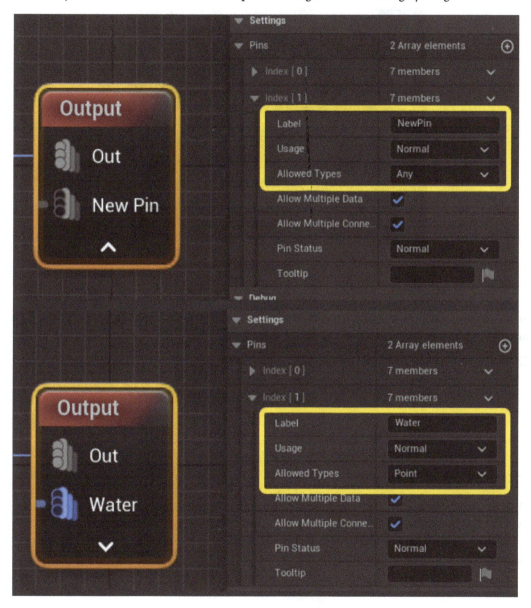

Figure 5.63 – Labeling and converting the output type

12. Having made those adjustments, let's now transition and reconfigure the connection from the **Out** to the **Water** output.

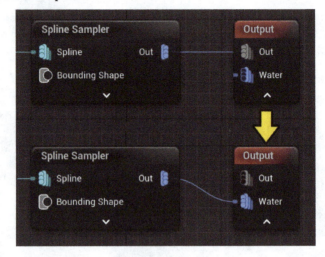

Figure 5.64 – Switching from the Output to the Water node

You have successfully finished creating the PCG graph for river water flow and have acquired knowledge on how to label and employ classes to select any blueprint element from Unreal's library, enabling its integration with any PCG graph.

To summarise, the role of the PCG graph is to facilitate the natural flow of a river across the terrain while ensuring it does not interfere with other PCG systems. This functionality is key in creating a watercourse that crosses seamlessly through a forested environment.

We are done here, so we may now close the PCG_Water graph. In the next section, we will go back to the PCG_Forest graph and start adding new PCG graphs inside it!

Integrating PCG forest

This section represents the core component of the activity, where our objective is to integrate all the PCG graphs into a singular, comprehensive PCG graph: PCG forest. You have constructed PCG graphs with the necessary functionality that allows them to be seamlessly combined and utilized for various applications.

In this scenario, we will place all the graphs and establish connections between the PCG graphs using the appropriate nodes:

1. Open the PCG forest. In the Content folder inside the PCG folder, select PCG_Water, then drag and drop the PCG_Water graph inside the PCG forest graph editor space.

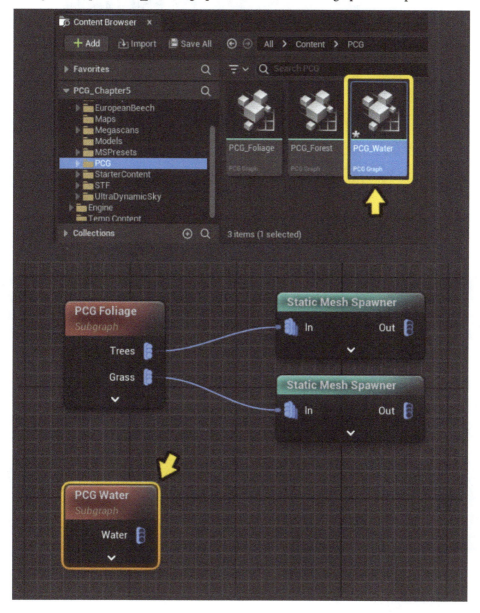

Figure 5.65 – Adding the PCG_Water subgraph to the PCG_Forest graph

2.   This will be an interesting part of the tutorial because we will use the nodes to connect one PCG graph to another. In this case, add the **Difference** node and connect the `PCG_Water` input with the **Differences** input (*Figure 5.66*).

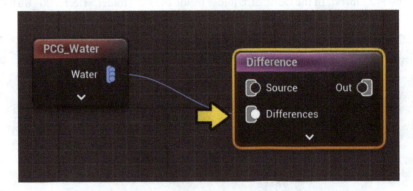

Figure 5.66 – Connecting the PCG_Water subgraph to the Difference node

3.   The purpose of adding the **Difference** node is to simply let the river pass through the foliage. In this case, grass will be moved apart to let the river flow across the landscape. For this reason, let's connect the grass node from `PCG_Foliage` to the **Source** input inside the **Difference** node.

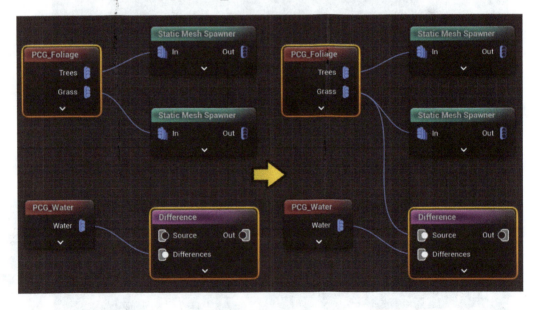

Figure 5.67 – Connecting the Grass output to the Difference Source output node

4. At the moment, grass won't be spawning as we would like it to. To complete this part, let's disconnect the **Static Mesh Spawner** node from the **Grass** input and connect the **Difference** node output to the **Static Mesh Spawner** node.

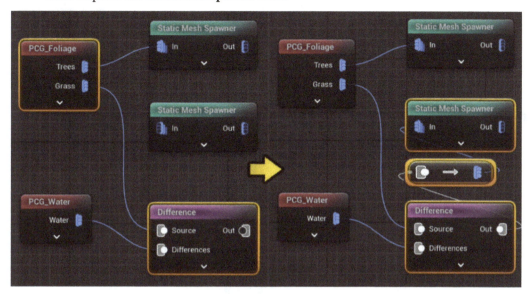

Figure 5.68 – Connecting the Difference node to the Static Mesh Spawner node

5. With the **Difference** node selected, navigate to the **Settings** panel on the right-hand side and alter the **Density Function** value from **Minimum** to **Binary**. This adjustment will clear the path through the grass, enabling the river to seamlessly flow through the landscape

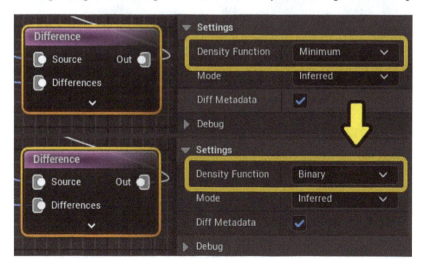

Figure 5.69 – Changing Difference's Density Function value from Minimum to Binary

6. Let's head back to the viewport and examine what changes it made after you applied those modifications. Move the water river body and place it on top of the grass and see what happens. You will notice that a tiny pathway has formed in the grass field. Let's make it a little bigger:

Figure 5.70 – Testing the water river body with the BP_Forest

7. Between the **PCG_Water** node and the **Difference** node, add the **Bounds Modifier** node and connect it.

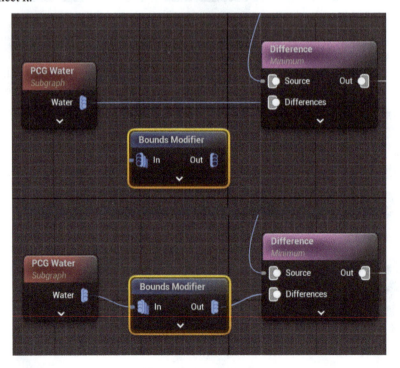

Figure 5.71 – Connecting Bounds Modifier to PCG_Water and the Difference node

8. With the **Bounds Modifier** node selected, on the right-hand side, change the **Minimum** and **Maximum** values as shown in *Figure 5.72*. This will move the grass to form more spaces for the river to flow.

Figure 5.72 – Changing the Bounds modifier values from 1.0 to 4.0 at the Y axis

9. Let's return to the main viewport to observe the outcomes. You'll notice that there is increased space available for the river, and wherever you relocate the river body, it will carve a new path while allowing the grass to regrow in areas previously occupied by water. Let's repeat the same process but for the trees!

Figure 5.73 – The Bounds Modifier node helping to remove the unnecessary foliage around the river

10. Head back to your **PCG_Forest** graph and copy the following nodes from the diagram below: **Bound Modifier**, and **Difference nodes**. Paste those nodes next to each other. Connect and rearrange them accordingly similarly to what we have achieved with the grass but this time for **Trees**!

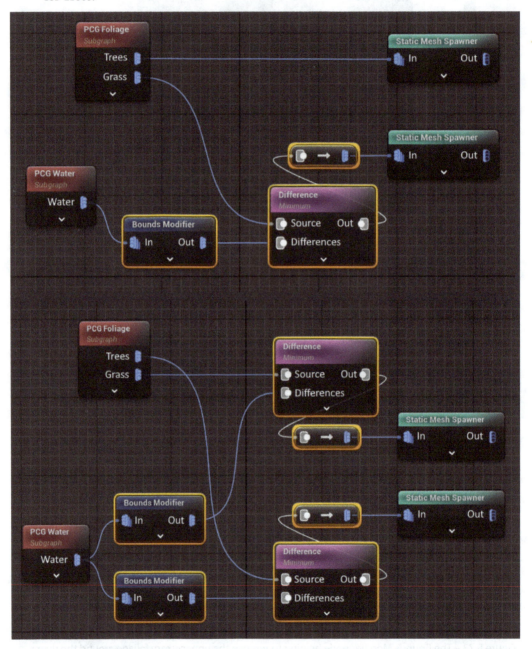

Figure 5.74 – Overview of the PCG graph that works with foliage and the water actor

11. Let's take a look at the main viewport and observe how it all turns out on the scene. For better results, you may want to adjust the **Bounds Modifier** node for the trees and adjust its size for bound minimum and maximum. The results are way better now!

Figure 5.75 – The river is crossing through the foliage

> **Note**
>
> To maximize the benefits of this setup, it's advisable to leverage the nodes within the blueprint. Starting with the **Bounds Modifier, Union,** and **Self Pruning** nodes to adjust values can significantly impact how models are spawned across the surface. This approach allows for observing the resulting changes in model distribution.

As we end this tutorial, revisiting the PCG graphs for one more iteration could be highly beneficial. Implementing additional refinements will further improve the quality of model distribution across the surface. In the upcoming section, we will review several nodes again and explore how adjusting the values can lead to better outcomes.

Making adjustments for better outcomes

In this section, we'll begin by altering the PCG graphs. This time, we will take a look at the PCG_ Foliage graph. Within this graph, the crucial part involves randomizing the transformation of the spawned foliage models on the surface. This step is key to achieving a more realistic appearance, as it helps to diminish the uniformity in the placement of grass and trees, thereby enhancing the natural diversification and realism of the scene:

1. Open your `PCG_Foliage` graph inside the **Content** folder. In the graph, select the **Transform Points** node that connects with the **Trees** output node.

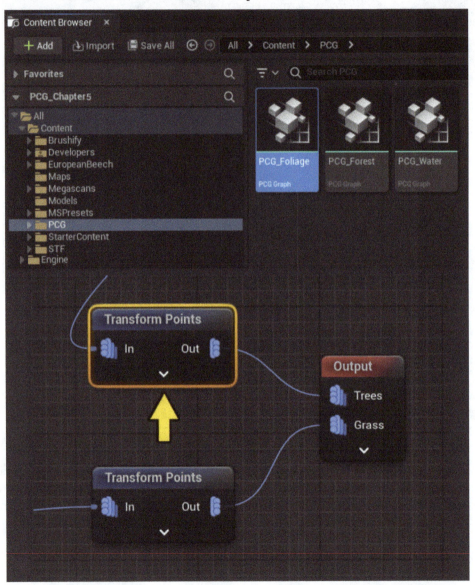

Figure 5.76 – Adding the Transform Points node to adjust the transformation of the trees

2. With this **Transform Points** node selected, under the **Settings** panel on the right-hand side, change the values for the **Offset**, **Rotation**, and **Scale values** you can find the **Transform Points** table with all the values in *Table 5.1*). This will help to randomize the trees' transformation. Enable **Absolute Rotation**.

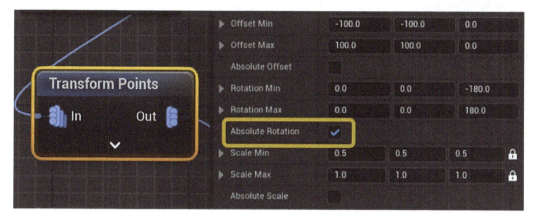

Figure 5.77 – Enabling Absolute Rotation inside the Transform Points node

In the diagram below, the numbers indicate the values to be added in the **Transform Points** settings panel.

Transform Points	X	Y	Z
Offset Min	-100.0	-100.0	0.0
Offset Max	100.0	100.0	0.0
Rotation Min	0.0	0.0	-180.0
Rotation Max	0.0	0.0	180.0
Scale Min	0.5	0.5	0.5
Scale Max	1.0	1.0	1.0

Table 5.1 – The values for the Transform Points settings details panel

3. Let's repeat this process but this time we will update the **Transform** points for the **Grass** output node. You will find the table with the transform points in *Figure 5.78*:

Figure 5.78 – Transform Points value inside the settings panel

In the table that follows, the numbers indicate the values to be added in the **Transform Points** settings panel.

Transform Points	X	Y	Z
Offset Min	-20.0	-20.0	0.0
Offset Max	20.0	20.0	0.0
Rotation Min	0.0	0.0	-180.0
Rotation Max	0.0	0.0	180.0
Scale Min	0.8	0.8	0.8
Scale Max	1.2	1.2	1.2

Table 5.2 – The values for the Transform Points settings details panel

4. Upon your return to the main viewport, you will find that all the foliage, grass, and tree models have been transformed in random directions.

Figure 5.79 – The final outcome of the BP_Forest actor blueprint

After implementing all the modifications, begin exploring the environment and experiment by moving the **WaterBodyRiver**. This will allow you to observe the impact of the changes on the scene and see how different arrangements can affect the overall appearance and realism of the environment.

Summary

In this chapter, you acquired important knowledge on utilizing spline control to manage the PCG graph. You've learned how to configure your actor blueprint to operate with the spline component, and how to merge various PCG graphs into a primary PCG graph for use as a component within the actor blueprint.

This approach is highly effective while working with a number of objects that are generated across the landscape, offering the flexibility to adjust asset numbers in real time as you work within the viewport.

In the upcoming chapter, we will explore landscape tools and the application of landscape materials to procedurally generate content across the terrain. This promises to be an engaging way to interact with the landscape, as it allows for painting while simultaneously creating your PCG volumes in accordance with the areas you're working on. This technique offers a creative and dynamic approach to landscape design, seamlessly blending manual customization with automated generation for enhanced environmental detailing.

Get This Book's PDF Version and Exclusive Extras

UNLOCK NOW

Scan the QR code (or go to packtpub.com/unlock). Search for this book by name, confirm the edition, and then follow the steps on the page.

Note: Keep your invoice handy. Purchases made directly from Packt don't require an invoice.

6

Building a PCG Graph with Landscape Materials

This chapter stands out as an exhilarating one to engage with. Following *Chapter 5*, where you delved deeper into define PCG graph and its flexible integration with other PCG tools, you achieved significant advancements by mastering the intricate functionalities of the PCG framework itself. This chapter presents a unique opportunity for you to create your own tool, capable of randomly placing cabin houses and procedurally generating foliage around architectural structures, including bodies of water such as rivers!

This chapter introduces a novel aspect of **PCG (Procedural Content Generation)** by leveraging Landscape Materials to control the placement of static meshes throughout the landscape. This approach presents a unique method for guiding the PCG system, concentrating specifically on regions painted with designated material setups. Building on the knowledge from *Chapter 5*, where we explored the use of a Spline component, this chapter will integrate that mechanism with landscape materials.

By doing so, we harness both the spline controller's path-defining capabilities and the landscape materials' environmental cues to finely tune the generation and placement of static meshes. This integration allows for a sophisticated, material-aware PCG system that dynamically adapts to the landscape's painted features, ensuring coherent and contextually appropriate content generation.

This chapter is structured into sections covering Landscape Materials, the PCG graph correlating with Landscape materials, and creating an Actor Blueprint for dynamic landscape interactions. Top of Form You'll become skilled in new techniques using PCG, enabling you to handle various landscape materials and integrate them with PCG graphs effectively.

Furthermore, this chapter's contents will be accessible via the GitHub link provided in the *Technical requirements* section.

This chapter focuses on the following key areas:

- Getting familiar with the Landscape

- Setting up the Landscape Pro 2.0 Auto-Generated Material assets, and the textures for the landscape material

- Developing and configuring an Actor Blueprint with spline control

- Constructing and integrating the PCG graph into the Actor Blueprint

This hands-on tutorial will equip you with the ability to shape your PCG volume precisely and generate models procedurally across the landscape. Upon finishing this chapter, you'll have acquired new techniques and valuable insights into developing a PCG tool, controllable through spline controllers and Landscape materials.

Let's get started!

Technical requirements

You will need the following to complete this chapter:

- A good computer that can run a PCG framework. Ideally, it should have a multi-core CPU (AMD Ryzen 7/9, Intel i7/i9), GPU (NVIDIA RTX, AMD Radeon RX with 8+ GB VRAM), and at least 16 GB RAM.

- Basic knowledge of Blueprint creation in Unreal Engine 5. The chapter is designed to be accessible to those with even a fundamental understanding of blueprint development.

- The template project comes without the assets and you will need to download the Marketplace. Make sure to download the Landscape Pro 2.0 Auto-Generated Material project files from the Unreal Marketplace. All of those assets are free of charge.

The Unreal Engine version used in this chapter is version 5.4. The PCG tool was introduced with Unreal Engine 5.2 but some nodes got deprecated in Unreal Engine 5.4, hence the latest version of Unreal Engine is the most beneficial for this chapter.

For this project, we will use the template project that is available in the GitHub repository. You can download the template from this GitHub link:

`https://github.com/PacktPublishing/Procedural-Content-Generation-with-Unreal-Engine-5`.

The code in action video for the chapter can be found at `https://packt.link/61it0`

Getting familiar with models

In this task, we will utilize the existing assets from the Landscape Pro 2.0 Auto-Generated Material package. We'll choose the tree models from the Content folder to enrich our example and prevent using identical assets across the book:

1. Navigate to STF -> **Pack03-LandscapePro** -> **Environment** -> **Foliage** -> **GreenTrees** and search for the two tree models called **SM_dead-tree05** and **SM_green-tree01**:

Figure 6.1 – Tree assets, SM_dead-tree05 and SM_green-tree01

2. Navigate to STF -> **Pack03-LandscapePro** -> **Environment** -> **Foliage** -> **Bush** and search for the **SM_bush01** asset:

Figure 6.2 – Bush asset, SM_bush01

3. Navigate to STF -> **Pack03-LandscapePro** -> **Environment** -> **Foliage** -> Clover and search for the **SM_clover01** asset:

Figure 6.3 – Grass asset, SM_clover01

4. The next step is to turn on **Nanite** mode for each mesh. With those models selected, right-click on one of the models and tick the **Nanite** checkbox at the top of the menu:

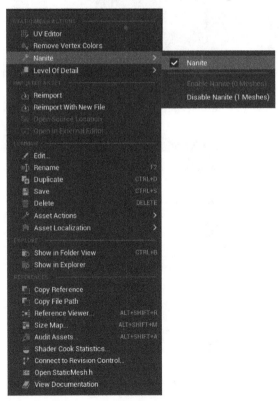

Figure 6.4 – Enabling Nanite mode for every foliage model

We are done with this section, so let's move on to more practical fun. In the next section, we will produce a Landscape material that will be used for this example and we will enable it to interact with the procedural content within the PCG graph.

Landscape Material

In this section, we'll delve into the creation of a specific type of material: Landscape Material. Landscape materials are highly effective and versatile, making them an excellent choice for various environment-creation tools, especially when integrated with a PCG graph. Therefore, we'll dedicate this section to crafting a material that is sufficiently straightforward to be compatible with PCG tools. Follow these steps:

1. Inside the `Content -> Materials` folder, right-click and add the Material and rename it `M_LandscapePCG`:

Figure 6.5 – Creating a new Material and naming it M_LandscapePCG

2. Double-click on the newly created material so we can start adding the right textures for this material. You will find the texture under `Content -> Textures`:

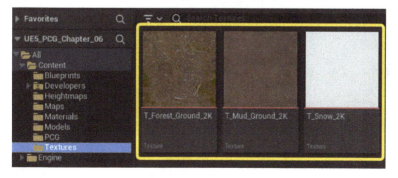

Figure 6.6 – Texture location within the project

Let's organize the textures in a column, making sure each one connects to the right input on the Material.

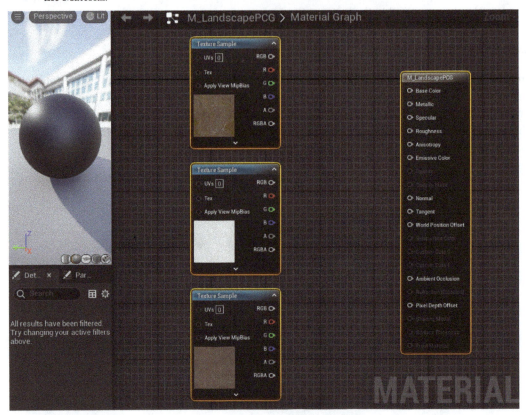

Figure 6.7 – Textures set up inside the material graph

3. The next step involves right-clicking within the **Material Graph** area and searching for LandscapeLayerBlend. This node allows the integration of the textures and nodes with the layers found within the **Landscape** tool:

Figure 6.8 – Adding a LandscapeLayerBlend node

4. We need to allow the textures to be drawn onto the landscape surface using **LandscapeLayerBlend**. Select the **LandscapeLayerBlend** node. Let's start adding elements to each array accordingly. In each **Index** section, change **Layer Name** to Ground, Mud, and Snow, respectively:

Figure 6.9 – Adding and renaming layers for the LandscapeLayerBlend node

5.   Once all the layers are configured, the next step is to link each texture to its respective inputs on the **LandscapeLayerBlend** node. The primary **LandscapeLayerBlend** node should then be connected to the **Base Color** input on the Material attributes, as shown here:

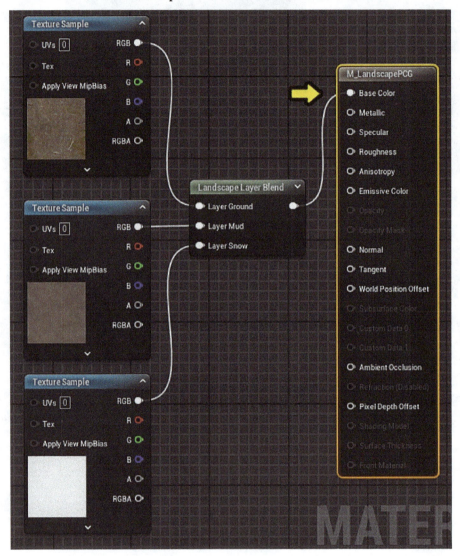

Figure 6.10 – Connecting to the Base Color

6. This will be enough to work with your PCG graph setup, but it will be much better if we can configure and resize the texture size. For this, let's add the *TextureCoordinate*, *Constant*, and **Multiply** node! The following figure shows the correct structure:

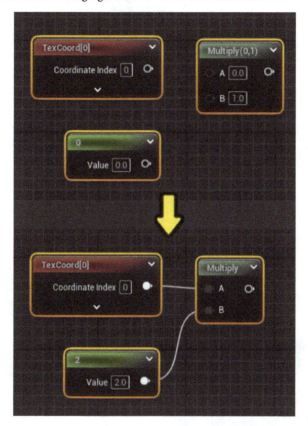

Figure 6.11 – Adding variables to control the texture size

7. Let's connect the nodes connection shown in *Figure 6.12* to each **Texture Sample** node's **UVs** input. This is enough for this part of the tutorial and you can close your landscape material.

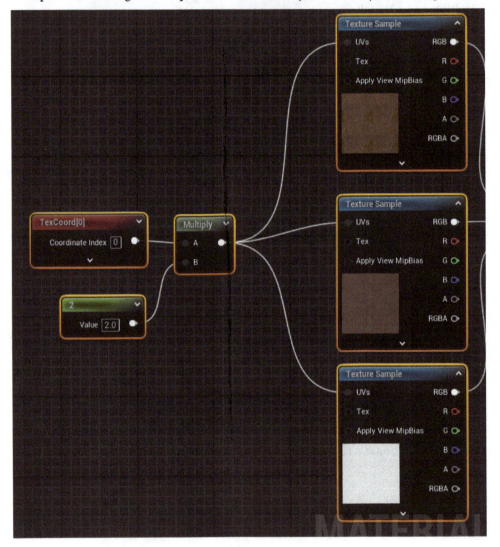

Figure 6.12 – Overview of the variables' connection to all the textures

8. Your material will look too shiny, and in this case, I will reduce the **Specular** property to zero. Let's connect another *Constant* with a zero value but, this time, to the **Specular** input node:

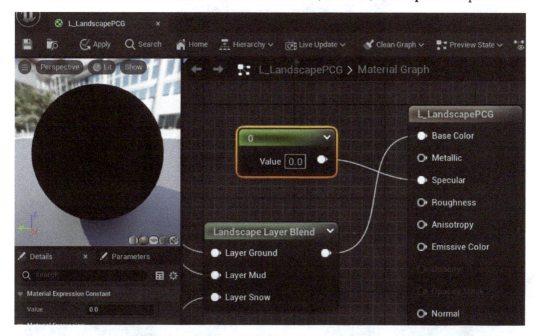

Figure 6.13 – Turning off the Specular property to remove the shininess

We just completed the material section, which was quite straightforward and it didn't require too much attention. In the next section, we will look at assigning your Landscape material and we will use layers to paint the landscape with the different textures.

Landscape tool

In this section, we will dig into painting our landscape using the **Landscape** tool. Thanks to the previously created material, we can achieve a very simple landscape that will be suitable for this example. Without further ado, let's get started on this exciting exercise:

1. Navigate to the Outliner window on the right-hand side and select the **Landscape** asset. With the **Landscape** asset selected, Under the **Landscape** tab, assign your newly created material to the **Landscape** asset:

Figure 6.14 – Selecting the Landscape asset on the Outliner

2. With the material selected, drag and drop it into the **Landscape Material** slot inside the
    Landscape actor:

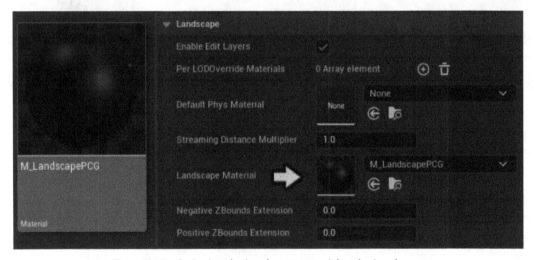

Figure 6.15 – Assigning the Landscape material to the Landscape

3. Your landscape will appear black because we haven't assigned the layers yet for each sub-material. To do that, select **Landscape Mode** in the top-left corner and then select **Paint**:

Figure 6.16 – Enabling Landscape Mode and selecting the Paint mode

4. Navigate down to **Target Layers** and, on the **Layers** tab, you will find three separate materials being shown on the panel. No layer has been assigned yet. To begin with, next to the first layer, click the plus (+) button and let's start adding Weight-Blended layers to each layer:

Figure 6.17 – Adding the Weight-Blended Layer information to each layer material

5. You will be prompted to save it in a default location within the Content folder. Therefore, save it as it is and repeat the same process with the other layers' information:

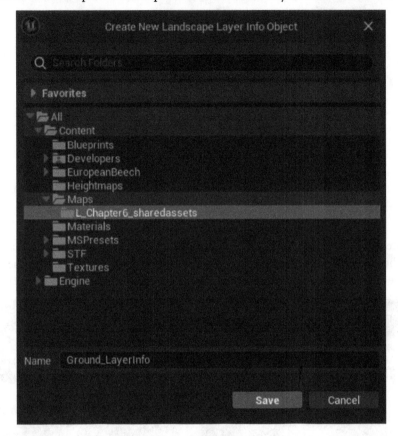

Figure 6.18 – Saving each layer to the default folder within the project

6. Once you've assigned the necessary layers info and are ready to begin painting the landscape, you may encounter an issue where your **Paint** tool appears red, accompanied by a message stating, **This layer has no layer info assigned yet**. This is a known glitch in Unreal, typically occurring when layer settings aren't established at the beginning of landscape creation. However, there's a workaround for this problem, which I'll detail in the following section.

Figure 6.19 – Missing a layer info message

7. On the **Edit Layers** tab on the left, right-click on **Layer** and select **Clear** -> **All**:

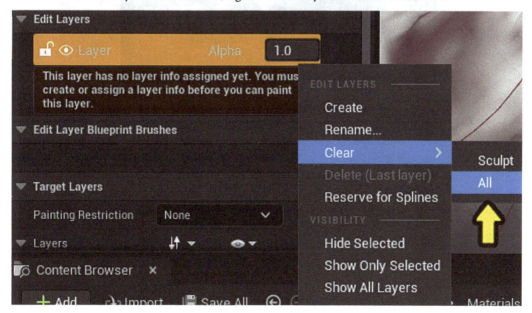

Figure 6.20 – Clearing all the layer's info

8. A message will pop up stating that the Layer content will be completely cleared. Press **Yes** to continue:

Figure 6.21 – Message to clear the layer's contained content

9. This action will clear the **Landscape** asset from the layer and your landscape will disappear from your editor. There is no reason to panic because we want to undo this action, so simply press *Ctrl + Z*, and your **Layer** info under **Edit Layers** will be fixed this time!

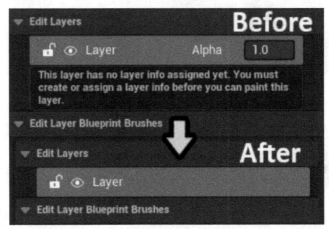

Figure 6.22 – Fixing the Edit Layer's layer

10. It's normal if your landscape currently seems to have a singular material with noticeable tiling repetition. This scenario is intentional for this exercise, aimed at demonstrating how you can merge PCG with the Landscape material. Feel free to paint the landscape as you wish. However, in this example, I'll guide you through my methodology. It's straightforward yet sufficiently illustrative to showcase the formation of the PCG graph later in the chapter.

Figure 6.23 – Painting the landscape with three color material

That concludes our section on the **Landscape** tool, providing us with a solid foundation to proceed with PCG creation. In the upcoming section, we'll delve into the Actor Blueprint before exploring the PCG graph.

Actor Blueprint

Continuing with our exercise, we've already accomplished two key steps: creating a Landscape material and painting the landscape to suit our needs. Now, we'll move on to preparing the Actor Blueprint, following the approach that we have used in previous chapters. Let's get started:

1. Inside `Content -> Blueprints`, right-click and choose **Blueprint Class**. Create a new **Actor** blueprint and name it `BP_PCGLandscape`:

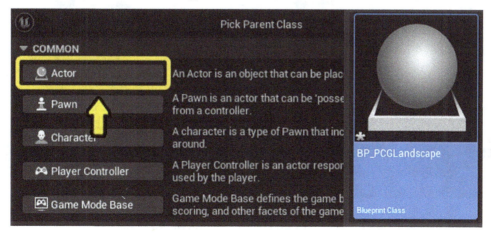

Figure 6.24 – Creating the Actor Blueprint

2. Open the **BP_PCGLandscape** Actor blueprint and, on the left-hand side under the **Components** tab, click the **+ Add** button to add the **Spline** and **PCG** components:

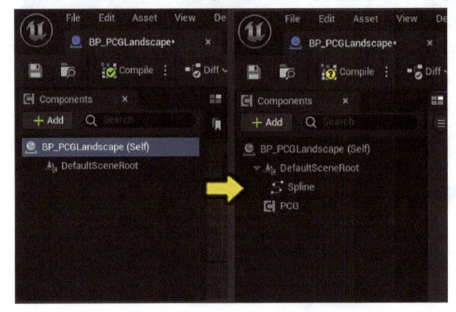

Figure 6.25 – Adding the Spline and PCG components

3. Select the **Spline** component and create the loop shape similarly to how we did in *Chapter 5*. This is the shape (*Figure 6.26*) we want to use for our exercise:

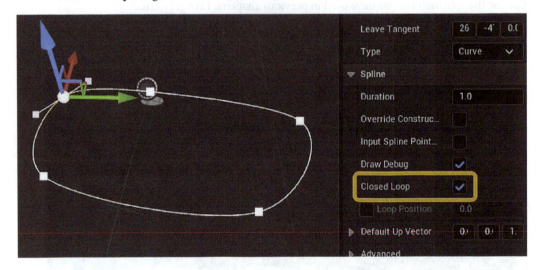

Figure 6.26 – Shaping the Spline controller and closing its loop

The next step involves adding some variables to the Actor blueprint, which will appear as Actor properties within the PCG graph. We'll introduce two **Float** variables and label them `TreesSpacing` and `PlantsSpacing`. Ensure these variables are exposed so that they are visible on the Details panel in the editor window. Just follow the example shown in the following figure:

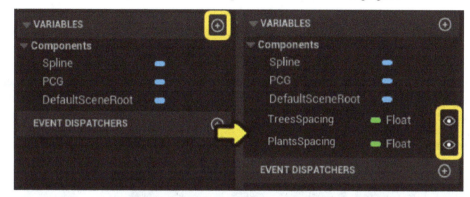

Figure 6.27 – Adding Float variables to control the spawning quantity of static meshes

Before proceeding to the next step, compile your actor blueprint. Adjust the default number values for both variables. Navigate to the **Details** panel on the right-hand side and configure the **Slider Range** and **Value Range** values for each variable to be between 50 and 2000. Set **Default Value** for each variable at 200 (*Figure 6.28*); otherwise, if the default number remains 0, it may cause some performance issues later within the PCG graph and it will crash your project.

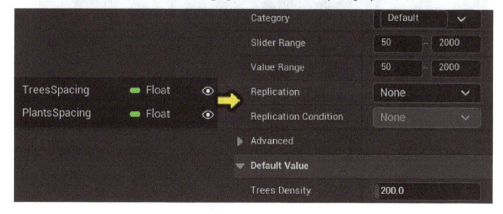

Figure 6.28 – Setting up the values for each variable

4.   You can now hit the **Compile** button and close your **BP_PCGLandscape** actor blueprint!

Having completed the *Actor Blueprint* section, you'll use this actor blueprint as a placeholder for the next section of this chapter, where we will delve deeply into the formation of the PCG graph and we will detail the process of constructing the graph to function effectively with your Actor Blueprint **Landscape** tool.

PCG graph

This section is pivotal in our exercise because we will utilize the PCG graph to project and spawn assets onto the Landscape plane. We'll introduce several new nodes designed to evenly distribute content and ensure that the assets align with the corresponding color assigned to the Landscape material. Without further delay, let's dive in!

1. In `Content -> PCG`, right-click and create a new **PCG Graph**. Rename your new PCG graph `PCG_Landscape`:

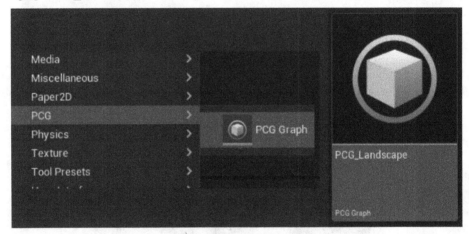

Figure 6.29 – Adding the PCG graph to the PCG folder

2. Open your **PCG_Landscape** graph and, inside it, start adding the following nodes: **Get Spline Data**, **Projection**, and **Spline Sampler**. Duplicate it to have two sets of the same nodes aligned vertically to each other:

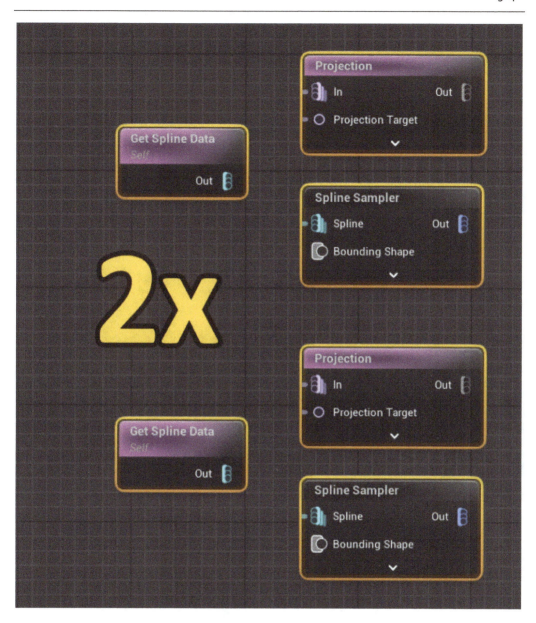

Figure 6.30 – Creating and duplicating the nodes graph

3. For both **Spline Sampler** nodes, let's use the following settings:

    - **Dimension**: **On Interior**

    - **Unbounded**: Checked

    - **Interior Sample Spacing**: 200

    This will provide the right setup to procedurally generate the static mesh within the interior space inside the **Spline** controller area:

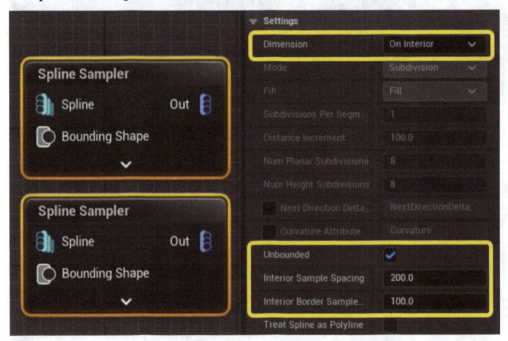

Figure 6.31 – Configuring the Spline Sampler nodes

4. Connect the nodes in the manner presented in the following figure. Connect the **Get Spline Data** node to the **Spline Sample** node and connect the **Spline Sampler** node to the **Projection** node. Repeat the same process with the duplicated nodes:

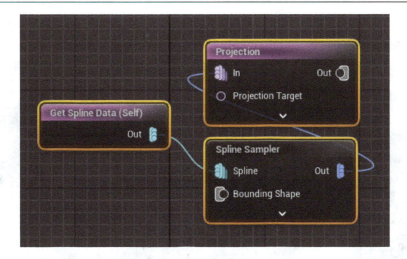

Figure 6.32 – Connecting all the nodes together

5. With this part completed, select the **Input** node and expand it by clicking the arrow-like button at the bottom of it. The node will show you all the inputs; we are going to focus on the **Landscape** input:

Figure 6.33 – Expanding the Input node and adding one input for the Landscape type

6. Connect the **Landscape** input of the **Input** node directly to **Projection Target** in the **Projection** node. Repeat the same step for the other nodes:

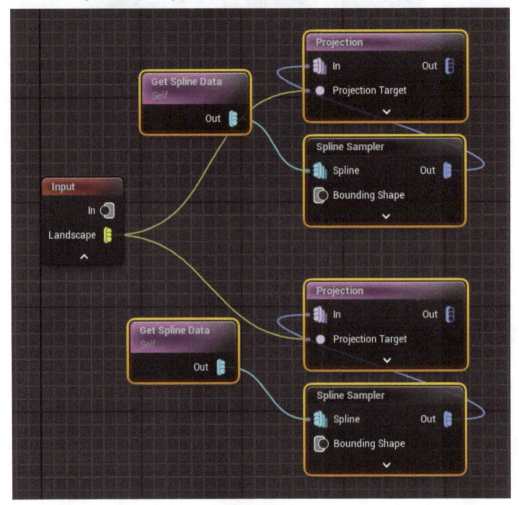

Figure 6.34 – Connecting the nodes to the Input node via the Landscape property

In summarizing this section of the exercise, we've incorporated **Spline** data into the **Projection** node, allowing projection onto the landscape through spline samplers. This use of two **Spline Sampler** nodes enhances the variety of mesh spawning effects. While a single **Projection** node could be used for this entire exercise, providing two illustrates how additional variety can be integrated into the existing graph.

Moving forward, we will introduce a special node within the graph, which we will explain shortly. This node will assist in filtering and determining the arrangement of the spawning meshes on the landscape in a more precise manner.

Attribute Filter

In the illustration below, the **Attribute Filter** node is essential for manipulating and refining point data based on specific criteria or conditions.

Figure 6.35 – The Attribute Filter node

This node is instrumental in spawning static meshes across the landscape, providing the ability to selectively process or exclude points from further operations within the graph based on defined rules or filters. In this scenario, **Inside Filter** and **Outside Filter** execute operations based on the conditions set for them. For instance, we can configure the static meshes to be procedurally generated at a specific location within the **Inside Filter** field. Conversely, **Outside Filter** will carry out the opposite operation, spawning objects outside its filter area.

With this knowledge in mind, let's continue with our exercise and explore how we can use this node to carry on within the PCG graph!

Mud layer

This time, we will use the **Mud** layer material to procedurally spawn the plants. For this, we need to prepare the condition that will enable the content to be generated on this section of the landscape. Follow these steps:

1. Inside the PCG Graph editor, right-click and search for the **Attribute Filter** node:

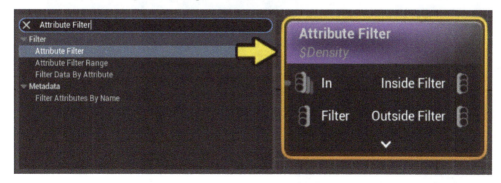

Figure 6.36 – Adding the Attribute Filter node to the graph

2. Let's connect the **Projection** node directly to the **In** input of the **Attribute Filter** node. For now, we will only focus on the top of the graph with the **Spline Sampler** node together with the **Projection** node. In this case, we will use the **Attribute Filter** node to create a condition that will determine whether the static meshes will be spawned on this section of the material or the other. The node will work similarly to the `if` statement and it will decide between both outputs coming from the **Attribute Filter** node: **Inside Filter** and **Outside Filter**.

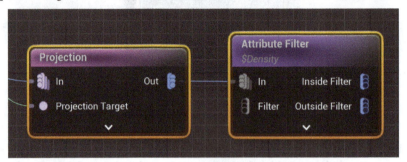

Figure 6.37 – Connecting the Projection node with the Attribute Filter node

3. Select the **Attribute Filter** node and, on the right-hand side under the **Settings** tab, set **Operator** to the >= condition. For **Target Attribute**, type in the word Mud. This condition will mean that the static meshes will only be procedurally generated if the area is greater than or equal to the area that covers the **Mud** material on the landscape:

Figure 6.38 – Configuring the Attribute Filter node

4. Before moving on to the next stage, let's remove the **Filter** input node on the **Attribute Filter** node and use the constant threshold instead. To do this, on the right-hand side, in the **Settings** panel, check the **Use Constant Threshold** box. This will allow us to expose the variable that will help to tweak and control the condition. Make sure to set **Type** to **Float** and set the value to 1.0:

Figure 6.39 – Enabling the settings within the Settings panel

5. With the **Attribute Filter** node prepared, let's add the **Transform Points** and **Static Mesh Spawner** nodes to the graph. Connect all the nodes together including the **Attribute Filter** node via the **Inside Filter** output.

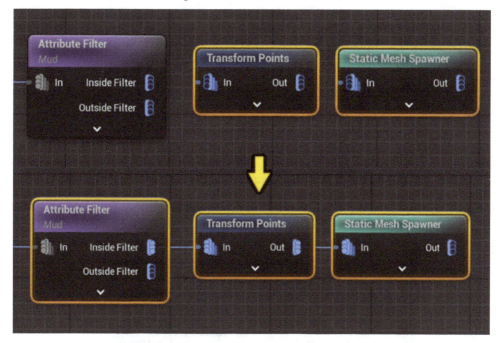

Figure 6.40 – Connecting all the nodes together

6.   Select the **Transform Points** node and, on the **Settings** panel, set up the settings as shown in *Figure 6.41*. You can also refer to the table that I prepared in *Table 6.1*, which

Figure 6.41 – Configuring the Transform Points node

7.   Following tabe shows the values that I used for the **Transform Points** node inside the graph:

Transform Points	X	Y	Z
Offset Min	-50.0	-50.0	0.0
Offset Max	0.0	0.0	0.0
Rotation Min	0.0	0.0	0.0
Rotation Max	0.0	0.0	0.0
Scale Min	1.0	1.0	1.0
Scale Max	1.0	1.0	1.0

Table 6.1 – Table representing the configuration setup for the Transform Points node

8.   Before we move on to the next node, let's test the current progress on your PCG tool! With the **Transform Points** node selected, hit the *D* key on your keyboard to activate the Debug mode. You should have a light blue octagonal icon activated in the top-left corner of the node:

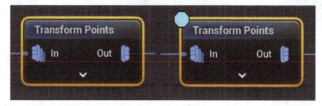

Figure 6.42 – Enabling the Debug mode inside the Transform Points node

9. Go to the `Content -> Blueprints` folder and open your **BP_PCGLandscape** actor blueprint. With the **PCG** component selected, on the right-hand side, in the **Instance** panel, add your **PCG_Landscape** graph:

Figure 6.43 – Adding the PCG graph inside the PCG component

10. The next step is to drag and drop the **BP_PCGLandscape** actor blueprint onto the scene, specifically where the muddy dark brown area is. Expand your spline controller to fit the entire **Mud** area and you should be able to see the white debug cubes being spawned on the landscape surface:

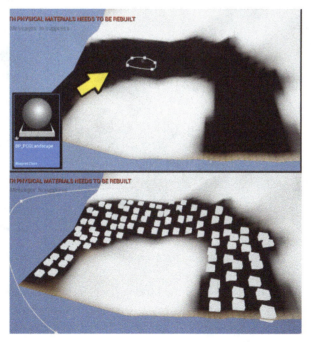

Figure 6.44 – Adding BP_PCGLandscape to the scene with the Debug mode enabled

11. After testing your tool, let's go back to our PCG graph and turn off Debug mode by simply selecting the **Transform Points** node and hitting the *D* key on the keyboard. Select the **Static Mesh Spawner** node and add one array under **Mesh Entries**. Search for **SM_clover01**:

Figure 6.45 – Adding static mesh inside the Static Mesh Spawner node

12. Let's go back to the editor window and preview the changes in real time. As you may have noticed, we now have plants being spawned on the **Mud** surface!

Figure 6.46 – Testing the spawning mesh on the Mud section of the material

13. To enhance the spawning of plants on the **Mud** surface, we should incorporate a property variable node. Let's add the **Get Actor Property** node and, in the **Settings** panel on the right, change the **Property Name** value to PlantsSpacing so that it matches the variable's name from the **BP_PCGLandscape** Actor Blueprint:

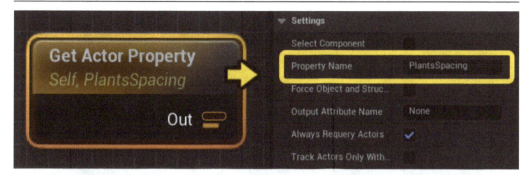

Figure 6.47 – Adding and configuring the Get Actor Property variable node

14. Click on the arrow at the bottom of the **Spline Sampler** node and expand all the variables inside the node. Search for **Interior Sample Spacing** and connect that input to the **Get Actor Property** node (*Figure 6.48*). When you have connected the **Get Actor Property** node to the right channel, click the arrow at the bottom of **Spline Sampler** again to close the entire node:

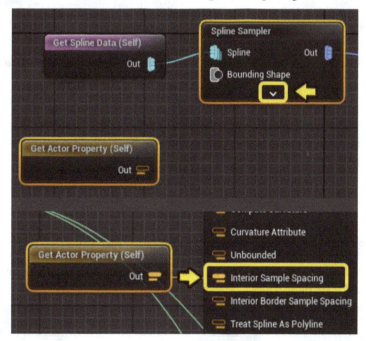

Figure 6.48 – Connecting the Get Actor Property node variable to Interior Sample Spacing

15. You now possess the capability to manage the procedural generation of plants exclusively on the **Mud** surface. Examine and test your **BP_PCGLandscape** actor blueprint through the Outliner window by scrolling down to a **Default** tab. Proceed to experiment with the PCG tool using various numbers, as demonstrated in the following example (*Figure 6.49*). This approach will greatly benefit you by offering substantial control and satisfaction through the **Settings** panel:

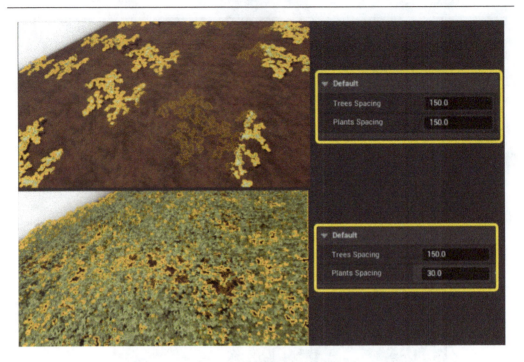

Figure 6.49 – Testing the plants property variable with the different values

In the next section, we will explore more point filters and add another condition to allow other static meshes to be spawned on the landscape but, this time, on the **Ground** and **Snow** layers of the Landscape material.

Ground layer

In this section, we'll employ the same strategy we used for the **Mud** texture material layer. We'll replicate the existing setup and switch out a few conditions with the new **Attribute Filter** node. Let's begin!

1. Inside the PCG graph, select the nodes that are shown in *Figure 6.50*. Duplicate the nodes (press *Ctrl + D*) and place those nodes as shown in the figure:

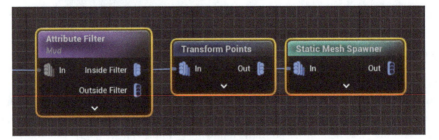

Figure 6.50 – Duplicating all the nodes

2. Connect the duplicated nodes to the second **Projection** node (*Figure 6.51*):

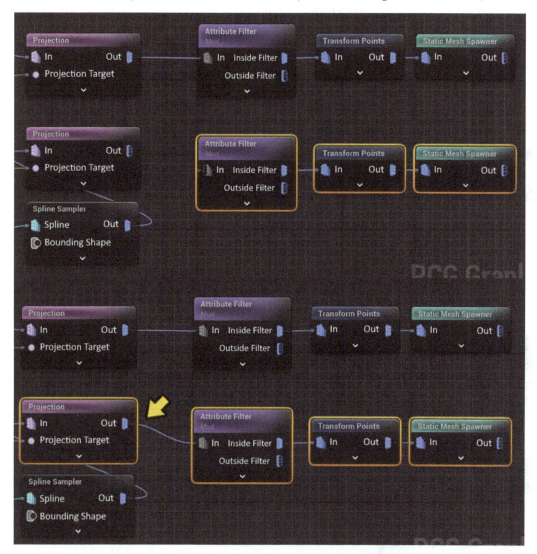

Figure 6.51 – Connecting the duplicated nodes to the second branch with the Projection node

3. Select the new **Attribute Filter** node that is connected to the second **Projection** node. This time, we need to set its **Target Attribute** name to Ground:

Figure 6.52 – Adding Ground as the Target Attribute name

4. At this point, we'll utilize the tree static meshes for the **Ground** layer material. Navigate to the **Static Mesh Spawner** node at the end of the graph and adjust the **Mesh Entries** values. For **Index [0]**, switch the **Static Mesh** model to **SM_green-tree01**. Then, add **Index [1]** and select the **SM_dead-tree05 Static Mesh** model for it:

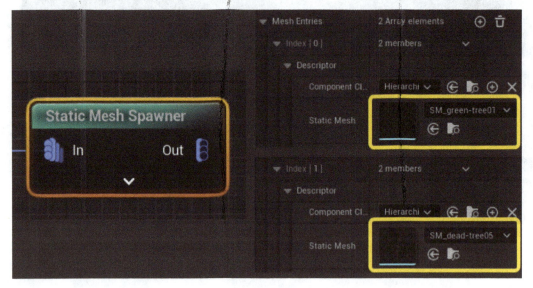

Figure 6.53 – Adding the tree models to Mesh Entries

5. Let's take a look at the viewport and see how it looks on the scene. It seems like we are getting the results that we wanted to get. In particular, all the trees are being spawned only at the **Ground** layer material location, just as we anticipated!

Figure 6.54 – Testing the Ground layer info material

You have achieved remarkable results in exploring the possibilities of using the material layers to procedurally generate the content on the landscape.

In the next section, we will use the **Snow** material to generate some bushes, which will add an extra dimension to the scene!

Snow layer

In this part of the exercise, we will replicate the process from the previous section by selecting the nodes we worked with and duplicating them again. This time, the duplication will be utilized specifically for conditions set to affect the **Snow** layer material surface on the Landscape. Let's get started:

1. Select the nodes that were used in the previous exercise, duplicate them using *Ctrl + D*, and place them below the original set. In this step, we will employ the **Outside Filter** output to differentiate the placement of content being spawned within the same area, but it will operate in a contrasting manner.

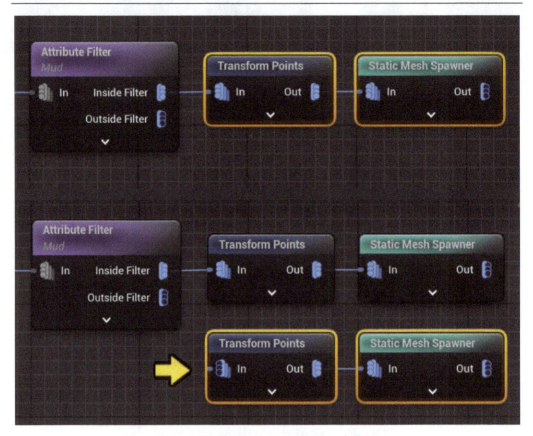

Figure 6.55 – Duplicating another set of nodes

2. Add another **Attribute Filter** node to the mix and let's change the condition to work only with the **Snow** layer surface of the landscape! *Figure 6.56* shows the configuration to use for this particular example:

Figure 6.56 – Configuring the Attribute Filter node's Settings panel

3. In this step, right next to the **Attribute Filter** node for **Mud**, let's connect all the nodes together to form a continuous chain:

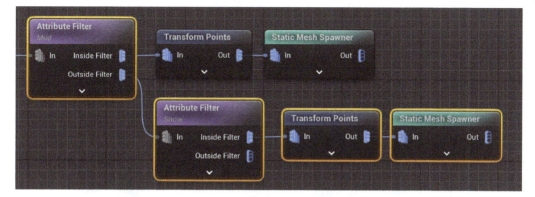

Figure 6.57 – Connecting a new Attribute Filter node to the extended nodes chain

4. Select the last node of the node network chain, which is a **Static Mesh Spawner** node. Delete one **Index** array from **Mesh Entries** and let's use one. For this one **Index** array, add the **SM_bush01 Static Mesh** model:

Figure 6.58 – Adding the SM_bush01 model to the Mesh Entries array

5. Let's examine the outcomes and review the modifications after we implemented another **Attribute Filter** branch to procedurally spawn the models onto the **Snow** layer surface. Let's go back to the viewport to check out the results:

Figure 6.59 – Testing the Snow layer info with the bush plants

6. It's definitely looking cooler and much better with the mix of different plants and trees inside this landscape. It will be worth adding another property to help control the number of trees and the little bushes. For this, we need to add a property variable node. Add the **Get Actor Property** node to the graph and set its **Property Name** value to TreesSpacing:

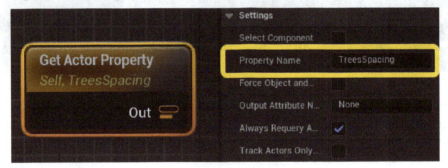

Figure 6.60 – Adding the Get Actor Property variable and setting the Property Name value to TreesSpacing

7. Expand the **Spline Sampler** node by clicking the arrow at the bottom of the node. Just like we did in the previous section, search for the **Interior Sample Spacing** input node and connect it with the **Get Actor Property** node:

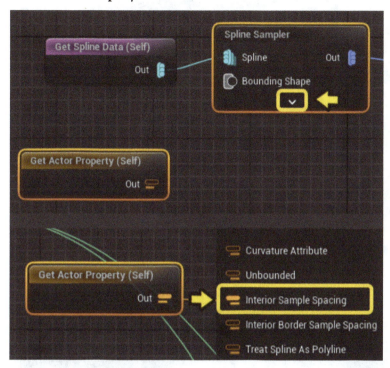

Figure 6.61 – Connecting the Get Actor Property node to Interior Sample Spacing

8. Everything is pretty much ready and we can test out our tool on the viewport. You can now select the **BP_PCGLandscape** actor blueprint and change the **Trees Spacing** values to compare the results:

Figure 6.62 – Testing the BP_PCGLandscape actor blueprint with the different values for each variable

You have just completed *Chapter 6* and learned how to efficiently create a PCG tool using only landscape material. This adds another valuable skill to your PCG knowledge library.

Summary

In this chapter, you've gained valuable insights into using spline control to influence the PCG graph's interaction with **Landscape** layer material information. This is built on the foundations laid in *Chapter 5*, where you learned how to configure your actor blueprint to work with the Spline component, blending the capabilities of **Landscape** material layers with the PCG graph. This method proved highly efficient when using **Landscape** tools and materials that necessitate PCG in a controlled environment. This method showed you how to create your own landscape materials and connect them to the right attributes within the PCG graph, allowing you to spawn different types of foliage based on the material's layer color.

In the next chapter, we'll dive deeper into the PCG Spline controller. You'll discover how to use splines to construct structures, equipping you with the skills to design simple buildings using logic nodes within the PCG graph. This method will be particularly helpful for creating buildings driven by specific nodes within the PCG framework. We'll also look at what's needed for a structure to take shape, using the **Transform Points** and **Spline Sampler** nodes to place static meshes along the spline, with the spline component built right into the Actor Blueprint.

Let's Build a Building Using the PCG Spline Controller

Welcome to *Chapter 7*! I hope you thoroughly enjoyed exploring landscape materials to distribute foliage procedurally in *Chapter 6*. Now, we're stepping up to a new challenge. In this chapter, you'll learn how to create buildings using PCG—yes, you heard me right, buildings! This is an excellent introduction for those eager to craft their own cities using the principles I'm about to share.

In this chapter, we'll dive into the methods of procedurally generating building structures, including walls that are complete with open gaps for windows. Following that, we'll develop an algorithm within the PCG graph that allows you to control and increase the number of floors. Get ready for an engaging journey into building your own urban building structure with PCG!

This chapter is structured into a primary section that addresses the walls, which are built using simple cube shapes. All tasks will be performed within the PCG graph, in collaboration with the Actor Blueprint, which facilitates the construction of the building and its associated functionalities.

You will gain proficiency in new PCG techniques, learning to manage different algorithms that assist in assembling and integrating static meshes into a coherent structure.

Upon finishing this chapter, you'll acquire new techniques and valuable insights into developing a PCG tool that's controllable through spline controllers. You'll also learn how to come up with structure designs based on the dimensions of the building.

This chapter focuses on the following key areas:

- Getting familiar with the custom wall model and Megascan assets
- Developing and configuring an Actor Blueprint with spline control
- Constructing and integrating a PCG graph into the Actor Blueprint
- Simplifying the graph structure together with the Actor Blueprint

Without any further ado, let's get started!

Technical requirements

You will need the following hardware and software to complete this chapter:

- A good computer that can run a PCG is a must. The computer must use a multi-core CPU (AMD Ryzen 7/9, Intel i7/i9), GPU (NVIDIA RTX, AMD Radeon RX with 8+ GB VRAM), and have at least 16 GB of RAM

- The template project includes modified Megascan assets. You can also download your favorite assets from Megascan, particularly the building structure assets from the Quixel Megascan library, at your convenience! All these assets are available free of charge.

- The Unreal Engine version that's used in this chapter is version UE 5.4. The PCG tool was introduced with this version.

For this project, we will use the template project, which is available in the GitHub repository. You can download its template from GitHub at `https://github.com/PacktPublishing/Procedural-Content-Generation-with-Unreal-Engine-5`.

The code in action video for the chapter can be found at `https://packt.link/8HSP3`

Quixel Megascan models

As we progress through this chapter, we'll utilize Megascan assets that I've prepped specifically for this exercise. We'll be working with assets that already have the correct dimensions because this exercise involves calculations to create building blocks that fit these dimensions. We will review this information later in the chapter. Now, let's examine the models that I have already prepared for this exercise:

1. Next, navigate to the `Models` folder. There, you will find a model called `SM_WallCube`.

Figure 7.1 – SM_WallCube static mesh model

As part of the model preparation process, we have to enable **Nanite** mode for each provided mesh in this example. Like we always do in this part of the exercise, right-click on each mesh and enable **Nanite** mode

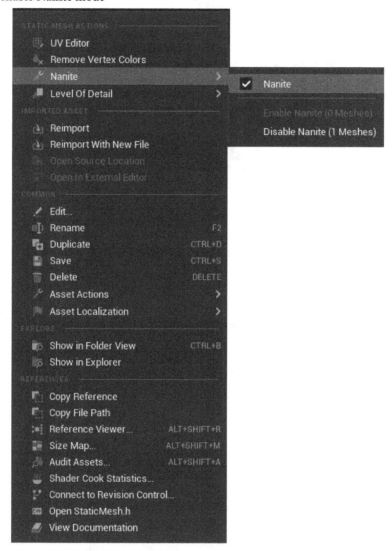

Figure 7.2 – Enabling Nanite mode

2. Next, we'll dive into the asset and blueprint needed for this example. After that, we'll create the actor blueprint, which will serve as the foundation for our building's structure.

Perhaps, while exploring the project, you might have already noticed that there is a PCG blueprint called `SetToPoints`. It is placed in the **PCG | Blueprint** folder.

Figure 7.3 – The SetToPoints PCG blueprint

I've customized the PCG blueprint just for you. It helps pinpoint locations on the first floor and lets us randomly place models such as doors.

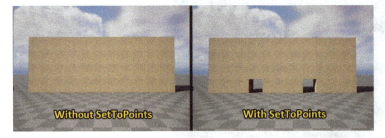

Figure 7.4 – Example of using the SetToPoints PCG blueprint ready for this chapter

We'll also use it to randomly set up spots for the windows. By creating this blueprint, I wanted to make things easier and quicker for future tasks, helping us work more efficiently and effectively.

In the following section, we will discuss why we selected these models and the importance of being cautious with their use due to their varying dimensions. In this exercise, we will construct the structure using the models provided specifically for this activity

Building structure Actor Blueprint setup

In this section of the chapter's tutorial, we'll take the time to create a crucial Actor blueprint that will be used to control and procedurally generate the building using the PCG component.

Let's start by creating your first Actor blueprint within the project and setting up the components you'll later control wfithin the PCG graph:

1.   Inside the Content Drawer in the **Blueprints** folder, right-click and search for **Blueprint Class**.

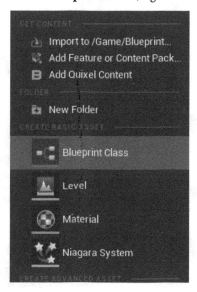

Figure 7.5 – Selecting Blueprint Class

2.   With **Blueprint Class** selected, inside the menu, choose the **Actor** class and rename it to BP_ BuildingStructure.

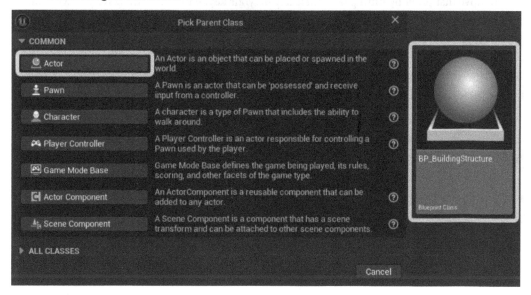

Figure 7.6 – Creating the BP_BuildingStructure Actor Blueprint

3. Now, double-click on your newly created `BP_BuildingStructure` actor blueprint. Then start adding the **Spline** and **PCG** components to your actor blueprint from the +**Add** menu.

Figure 7.7 – Adding the Spline and PCG components inside BP_BuildingStructure

4. Choose the **Spline** component. Inside the viewport, select the last point of the spline and remove that point by hitting the *Delete* key on your keyboard. This will leave you with only one point of the spline.

Figure 7.8 – Deleting the unnecessary spline point

5.   Now we have to create a square-shaped spline loop inside the viewport. Select the only remaining spline point inside the viewport and right-click above that spline point. The menu will show up. Select the **Spline Generation Panel** option.

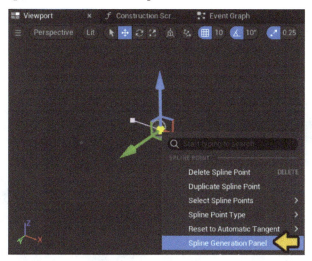

Figure 7.9 – Selecting the Spline Generation Panel option

6.   The **Spline Generation** menu will pop out and ask you what shape to choose. In this case, we will choose a **Rectangle** shape! A rectangular shape will make it easier to set up the layout for the new building, giving us a clear outline for all four walls. Make sure to change the shape parameters for **Length** and **Width** to 1000.0 units each.

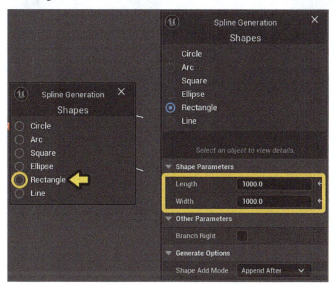

Figure 7.10 – Choosing a Rectangle shape with the adjustments for the Length and Width settings

> **Note**
>
> The latest version of Unreal Engine 5 introduces several new features, including the Spline Generation Shape tool. This tool is particularly useful for creating a variety of shapes quickly and easily using the spline component.

7. We need to close the loop with the **Closed Loop** checkbox on the right-hand side of the **Spline details** panel.

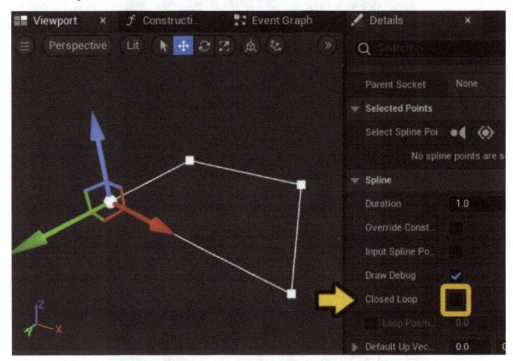

Figure 7.11 – Selecting the Closed Loop option to close the spline

8. After checking the Closed Loop box, you'll see that the selected point results in a twisted spline. This occurs because the last selected point reverts to its original setting, which is the Curve shape. To address this, select the fourth point on your spline and remove it by pressing the *Delete* key on your keyboard. Also, re-add the last point and change its location to 0 , 0 , 0 to get a perfect square.

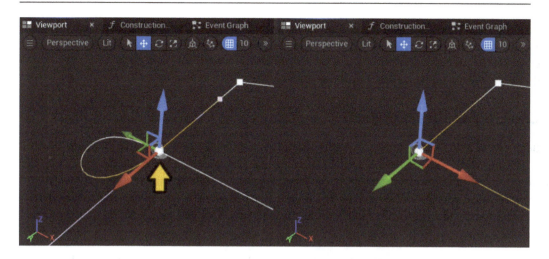

Figure 7.12 – Fixing the tangled loop at the beginning of the spline

9. The next step is crucial as it involves setting up variables that will control and manage the components of the PCG graph. This graph dictates elements such as the spacing between walls and the number of floors within the PCG volume. Let's create the necessary variables:

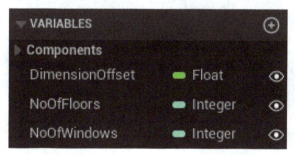

Figure 7.13 – Creating variables for DimensionOffset and NoOfFloors

- **Float variables**: Create a variable named `DimensionOffset`
- **Integer variables**: Create a variable named `NoOfFloors`

The values for each variable are presented in the diagram below. Make them all public:

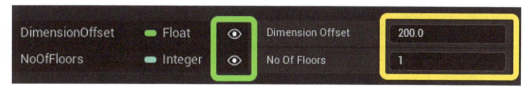

Figure 7.14 – The variables names and their values

Before we continue building our first graph, let's create a few macro functions. These will be used as our calculation panels and we will need them for later exercises. The macro functions will play a major role in calculating specific measurements that will be needed to calculate the pivots, as well as the offset of each building block static mesh position!

Creating an offset macro

Before we begin creating a macro, it's important to understand its purpose and relevance to our task. Initially, the walls we generate will be misaligned and positioned incorrectly. The offset macro (for more information, visit `https://dev.epicgames.com/documentation/en-us/unreal-engine/macros-in-unreal-engine`) plays a crucial role here, correcting these misplacements and ensuring the models are aligned properly. With this understanding, let's proceed with the process:

1. We will use the **Construction Script** tab, which is located next to the **Event Graph** tab in your actor blueprint. This is a designated area where you can interact with the blueprint from the editor's level without being in gameplay mode.

Figure 7.15 – Selecting the Construction Script tab

2. Inside the **Construction Script** graph, right-click in the graph space and search for two operators nodes: **Divide** and **Multiply**.

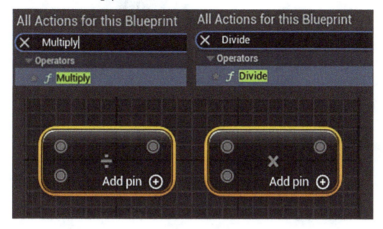

Figure 7.16 – Adding two operators

3. The next step is to drag out the **DimensionOffset** variable from the variables tab, connect it to the **Divide** node, and connect the rest of the nodes in the same manner as is shown in *Figure 7.17*:

Figure 7.17 – Constructing the node's structure to work with the Dimension Offset

4. As you can see, the **Divide** node changed to read the float values once we connected the float variable. Let's do the same with the **Multiply** node. With the **Multiply** node selected, right-click on the input pin where the 0 number is and convert it to **Float (single-precision)**. Make sure you convert all the remaining input and output pins to **Float (single precision)** on each node.

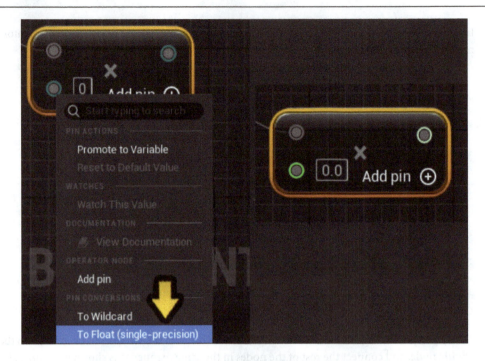

Figure 7.18 – Converting from Integer to To Float (single-precision)

5.   Connect all the nodes in the same manner as shown in *Figure 7.19*:

Figure 7.19 – Nodes structure for the Dimension Offset

6.   The next step is to convert our existing nodes into a macro function. Macro functions play a very important role in creating a functional node that can be reused to calculate things within the blueprint. In this case, instead of using default nodes that do specific calculations, you can create your own custom function that will suit your needs! Select the **Divide** and **Multiply** nodes, right-click on the selected nodes, and select **Collapse to Macro**.

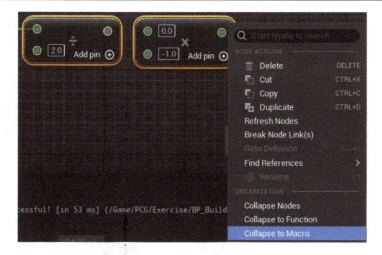

Figure 7.20 – Selecting the operator nodes and converting them to the macro function

7. A new macro function has been formed. Now, let's rename it to a suitable name: `Offsetter`. This node will play an important role in setting the position of the spawning walls and set its positions to match the middle point position of each static mesh.

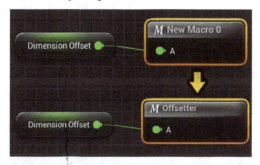

Figure 7.21 – Renaming the macro

Double-click on your newly created **Macro** node and let's add a few more outputs.

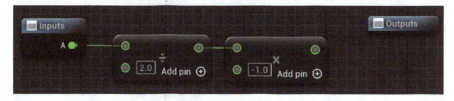

Figure 7.22 – Preparing inside the Offsetter macro

8. Click on the **Output** node and navigate to your right-hand side, to the **Inputs** and **Outputs** tabs. Click on the + icon and the **Offset**, **OffsetPosition**, and **ReverseDirection** float variables, just like in the figure that follows.

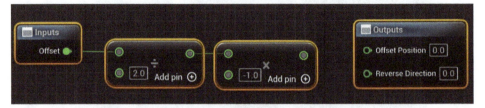

Figure 7.23 – Connecting the Offset input to the calculus float nodes

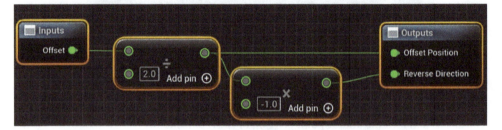

Figure 7.24 – Adding outputs properties and connecting to the calculus float nodes

9. Before advancing to the next step, we need to ensure that each pin is correctly connected to its corresponding variable, as outlined in the figure provided. This connection is critical for the calculations to function correctly, influencing the spawning of the static mesh. By meticulously aligning each offset's position, we ensure that the mesh spawns accurately and consistently within the scene. Follow these steps closely to set up your graph properly.

Figure 7.25 – Creating Inputs and Outputs float variables

The **Offsetter** macro is now created inside the graph editor.

Figure 7.26 – Offsetter Macro node

The **Offsetter** macro is now ready for use. Great! Now that we have a macro function ready, we're well-equipped to proceed with our project. Next, we'll dive into creating a PCG graph for the building structure.

In the upcoming section, we'll explore the network of connection nodes that will facilitate the construction of the building from the ground up. This step is crucial as it will lay the foundation for how the entire structure comes together dynamically. Let's move forward and start building our PCG graph!

Building a PCG Graph

In this part, we will get into creating our main engine of this exercise. This will all be done inside the PCG graph! Let's create our PCG graph:

1. Inside the **PCG** folder, create your PCG graph. Rename it to `PCG_BuildingStructure`.

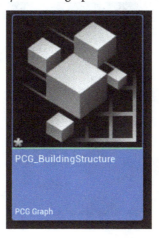

Figure 7.27 – Renaming the PCG graph to PCG_BuildingStructure

2. Before we move on to the next phase, save your PCG Graph and add it to the **PCG component** inside your BP_BuildingStructure actor blueprint in **Instance | Graph**.

Figure 7.28 – Assigning the PCG_BuildingStructure graph to the PCG component

3. Now go back to your content browser. Double-click to open your PCG_BuildingStructure PCG graph. Inside your graph, let's start adding a few nodes that will be used to create the walls around the building structure. These nodes are **Get Spline Data**, **Spline Sampler**, and **Get Actor Property**.

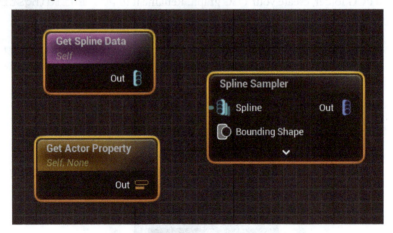

Figure 7.29 – Preparing the nodes inside the PCG graph

4. Before we start connecting nodes, let's examine the nodes and input a few values. Let's start with the most important one: the **Spline Sampler** node. Inside the **Spline Sampler** node, change the **Dimension** settings to **On Interior**. Make sure to check **Unbounded** at the bottom of the settings menu.

Figure 7.30 – Setting the Spline Sampler dimension to On Interior

Figure 7.31 – Enabling the Unbounded function inside the Spline Sampler settings

5. The next part is to set the right name to delegate the right variable inside the **Get Actor Property** node from the main `BP_BuildingStructure` blueprint. For this node, we will use the **Dimension Offset** float variable that we already used inside our actor blueprint. Select **Get Actor Property** and change its name under the **Property Name** to `DimensionOffset`.

Figure 7.32 – Adding DimensionOffest to the Property Name field

6. We have prepared the nodes and can now connect all of them accordingly. Make sure that **Get Actor Property** is connected to the **Interior Sample Spacing** input node inside the **Spline Sampler** node!

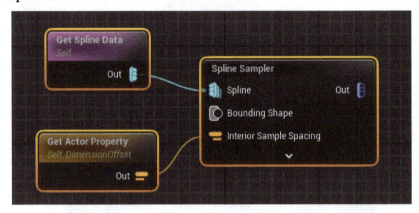

Figure 7.33 – Connecting the nodes with the Spline Sampler

7. The next phase of the exercise is to add and duplicate the **Transform Points** nodes four times. Connect them all to the **Spline Sampler** node.

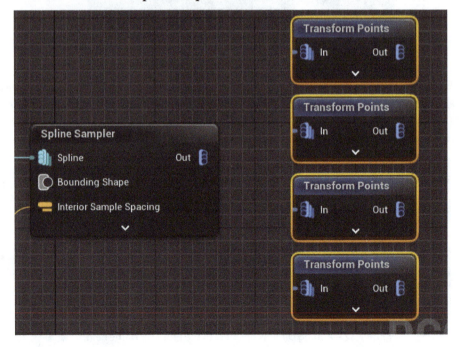

Figure 7.34 – Adding and preparing four Transform Points nodes

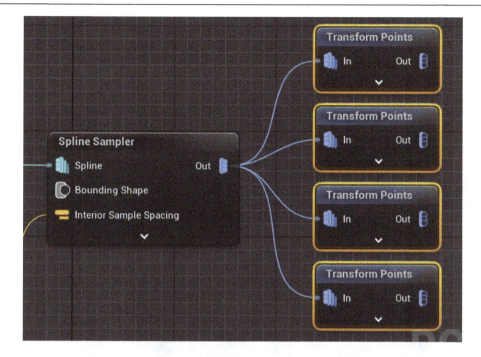

Figure 7.35 – Connecting all four Transform Points nodes to the Spline Sampler Output node

8.   We've correctly configured nodes together. Use the **Difference** node to eliminate any unnecessary interior walls; otherwise, it will create a maze-like interior inside our building structure. Add this node and duplicate it four times!

Figure 7.36 – Adding the Difference nodes to the PCG graph

9. Position the four **Difference** nodes adjacent to the **Transform Points** nodes. Refer to the illustration in *Figure 7.37* to see how to connect them. The purpose of this node structure is to identify the source, which is the wall in this instance, and differentiate it by removing any overlapping walls from the opposite side of the building structure.

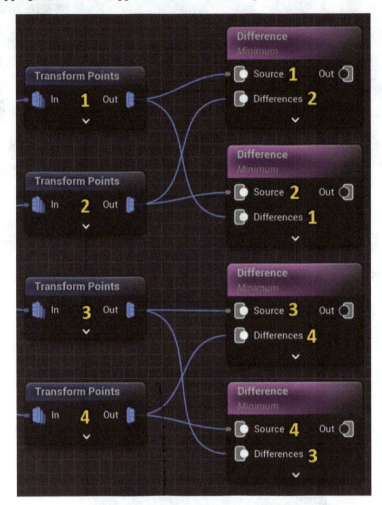

Figure 7.37 – Connect the numbered output and input nodes accordingly to ensure they match the correct connection points

10. We have completed one section of the whole graph. For now, let's debug our PCG graph by selecting all four **Transform Points** nodes. Simply activate the **Debug** mode by hitting the *D* key on your keyboard.

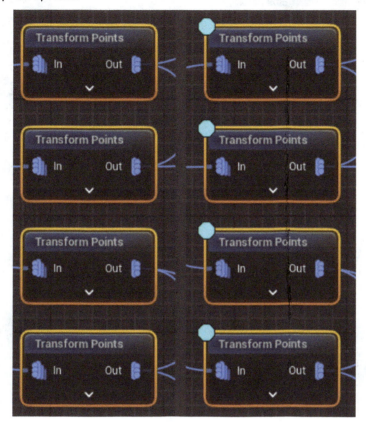

Figure 7.38 – Enabling the Debug mode to visualize the current progress

11. For each **Transform Points** node, adjust the **Min** and **Max** values on the Z-axis as shown in the following figure. This will enable the rotation of each wall for constructing the final structure.

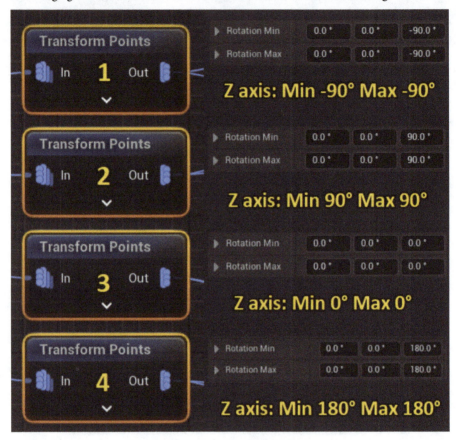

Figure 7.39 – Setting up the Z-Axis Rotation values for each Transform Points nodes

12. Go to your Content Browser and select your **BP_BuildingStructure** blueprint. For now, let's test the Actor blueprint by simply dragging and dropping the BP_BuildingStructure blueprint onto the scene! Based on your current setup, you can see that cubes are appearing on the grid formation. That's fine since we still have to add a few functionalities to our actor blueprint later in our exercise.

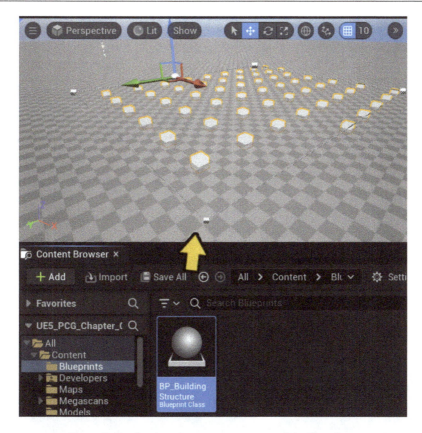

Figure 7.40 – Preview the current structure of the Transform
Points nodes to visualize the spawning positions

13. Let's go back to our PCG graph and add a few more variables that will be connected to the **Transform Points node**. At the top right of the PCG graph, there is a tab called **Graph Settings**. Click on it and it will open the menu on the right-hand side of the graph. Under the **Instance** tab, add four parameters.

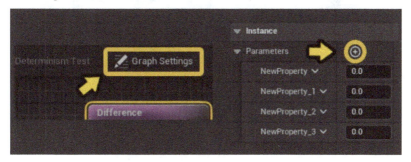

Figure 7.41 – Access the Graph Settings panel and add parameters inside the Instance tab

> **Note**
> Graph Settings were introduced as a new feature in the PCG graph with the release of versions 5.3 and 5.4 of Unreal Engine 5. This addition helps to create the internal variables that can communicate externally within the actor blueprint.

14. With each new parameter added to the list, go one by one and change them from the **Float** variable to **Vector**. Then rename them as follows: **WallWest**, **WallEast**, **WallNorth**, and **WallSouth**, just like in *Figures 7.42* and *7.43*.

Figure 7.42 – Change the parameters from Float to Vector and rename the variables accordingly

15. Each vector variable was renamed to its respective name.

Figure 7.43 – Rename the variables to WallWest, WallEast, WallNorth, and WallSouth

16. With those parameters added to the list, we can now use them inside our PCG graph. We will connect them individually to each **Transform Points** node corresponding to each direction offset. Right-click on the graph and search for all the parameters.

Figure 7.44 – Adding the parameters to the PCG graph

17. In this scenario, we will link the **Transform Points** nodes with the instance parameters from the **Graph Settings** tab. Select the arrow at the **Transform Points** node and connect it to both the **Offset Min** and **Offset Max** inputs.

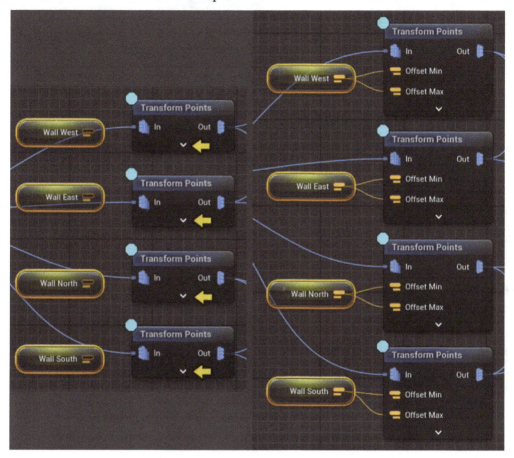

Figure 7.45 – Connect each parameter accordingly to match the Transform Points nodes

18. All **Transform Points** nodes pass through **Difference** nodes and must proceed to the meet point within the PCG graph structure. To consolidate all points into a unified structure, we should introduce a **Merge Points** node.

Figure 7.46 – Introducing the Merge Points node

19. Let's incorporate this node into the graph and connect it to all **Difference** nodes via the **To Point** link node, which we can be added inside the graph.

Figure 7.47 – Add the To Point sub-node to the graph and connect the Difference
Nodes via the To Points sub-node to the Merge Points node

20. To end this part of the exercise, let's add the **Transform Points** node and connect it with **the Merge Points** node. This will be used to fix the pivot offset position later in this chapter.

Figure 7.48 – Connect the Merge Points node to the Transform Points node

We've successfully organized and structured the PCG graph to spawn the walls. In the upcoming section, we'll revisit our BP_BuildingStructure actor blueprint and integrate additional functionalities that will enhance control over the spawned wall static meshes.

PCG parameters in Actor Blueprint

We have successfully set up the nodes within the PCG_BuildingStructure PCG graph, enabling us to construct a future wall structure in all four directions. Now, we'll begin integrating the PCG graph instance parameters and linking them to the appropriate macro functions that were developed in the previous section of this chapter.

To start, let's open the BP_BuildingStructure actor blueprint and begin adding the nodes required for this example:

1. Drag and drop the **PCG** component from the **Components** tab in your actor blueprint into the **Construction Script** graph space.

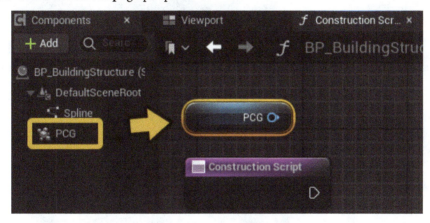

Figure 7.49 – Moving the PCG component from the Components window to the Construction Script graph

2. From the **PCG** component, drag out the pin and search for get **Graph Instance** node.

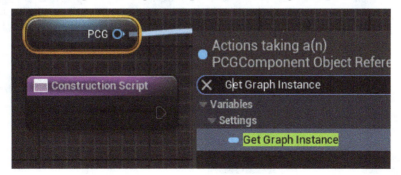

Figure 7.50 – Accessing the Get Graph Instance node

3.   From the **Get Graph Instance** node, drag out another pin. This time, search for the **Set Vector Parameter** node. Select this node and duplicate it four times. Make sure to connect the duplicated nodes back to the **Get Graph Instance** node!

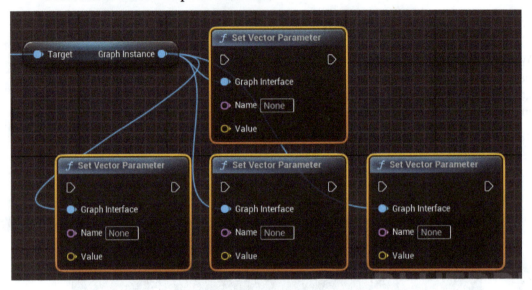

Figure 7.51 – Add the Set Vector Parameter nodes four times
and connect them to the Graph Instance node

4.   Let's add the names for each node inside the **Name** parameter as follows: WallWest, WallEast, WallNorth, and WallSouth.

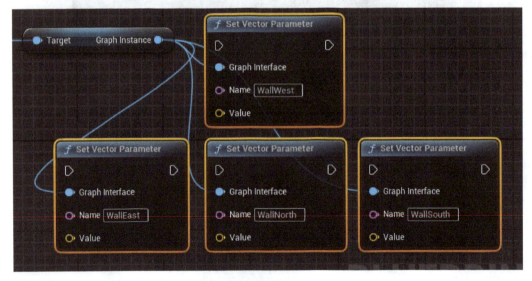

Figure 7.52 – Add the names to each Set Vector Parameter node

5. Since these are vector parameters, we need to access only the X and Y float input variables, which requires splitting the struct for each parameter node. To achieve this, right-click on the yellow symbol for the **Value** input on the node. Repeat this process for all other vector parameter nodes

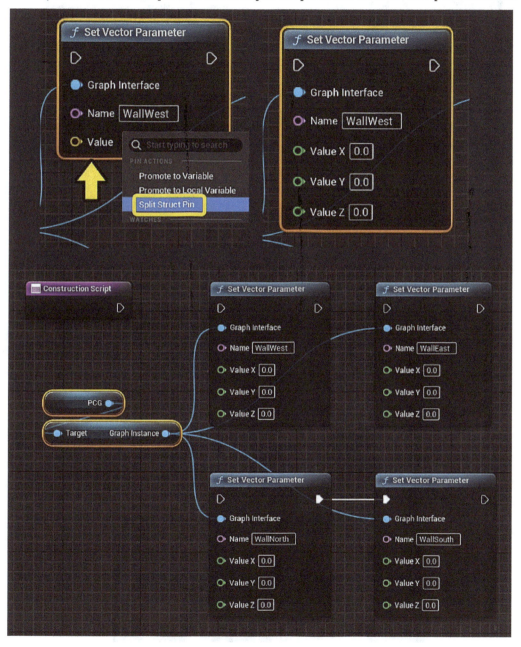

Figure 7.53 – Split Struct Pin from the Value input property

6. Drag and drop your **Offset** macro and link it with all the **Set Vector Parameter** nodes as needed! Follow the illustration as shown in *Figure 7.54*

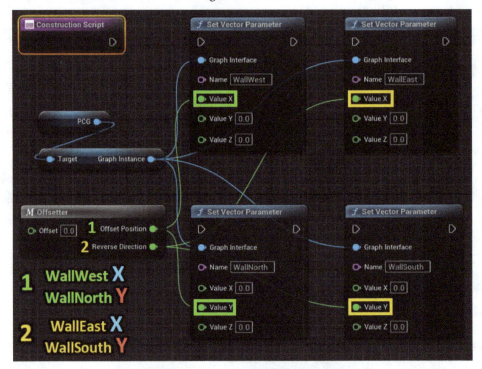

Figure 7.54 – Connect the Macro Offsetter output properties to each Set Vector Parameter's Values input property

7. The primary reason for using **Reverse Direction** is to replicate the walls while facing them in a different direction; otherwise, the newly generated walls would remain aligned along the same line of position as the original walls. Ensure that everything is connected as shown in the illustration and follow the instructions provided.

8. The next step is to finalize it by connecting the **Dimension Offset** variable to the **Offsetter** macro offset input pin!

Figure 7.55 – Connect the Dimension Offset float variable to the Offsetter macro

9. Connect all the execution pins from the vector parameter to the **Construction Script** tab. Compile your blueprint and save your progress.

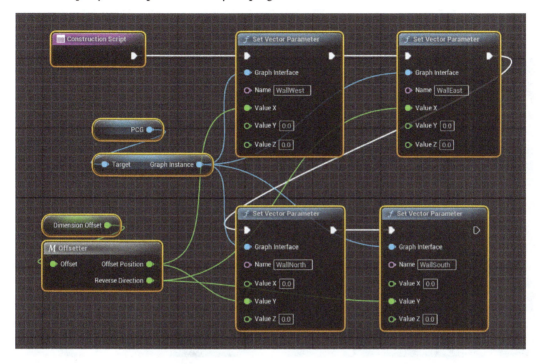

Figure 7.56 – Final look at the structure of the connected nodes inside the Construction Script graph

10. Head back to your PCG graph in PCG_BuildingStructure. Just hit the *D* key while you're on the Transform Points node that's connected to the Merge Points node to enable Debug mode.

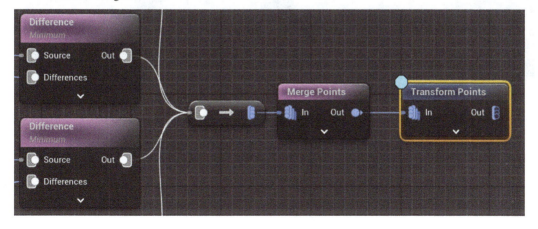

Figure 7.57 – Enable Debug mode to preview progress in the scene

11. As you might have noticed, the white cubes started to appear correctly on each side. This represents a foundation of the walls' formations on all four sides!

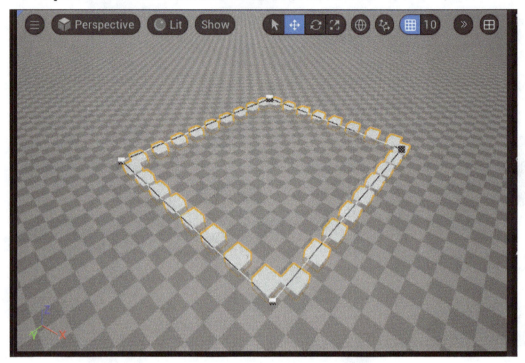

Figure 7.58 – The four wall sides are now being previewed in the scene

12. Let's turn off Debug mode on the **Transform Points** node. Also, let's add the **Static Mesh Spawner** node at the end of our graph inside the PCG graph!

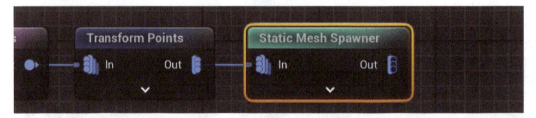

Figure 7.59 – Connect the Static Mesh Spawner node to the Transform Points node

13. With the **Static Mesh Spawner** node selected, navigate to the right-hand side panel. Under **Mesh Selector | Mesh Entries**, add an array element and insert the SM_WallCube static mesh.

Figure 7.60 – Add one mesh entry and assign the SM_WallCube static mesh

14. All four wall sides have been formed in our viewport scene! Of course, there are additional steps to take but this is a good indication that we are going in the right direction!

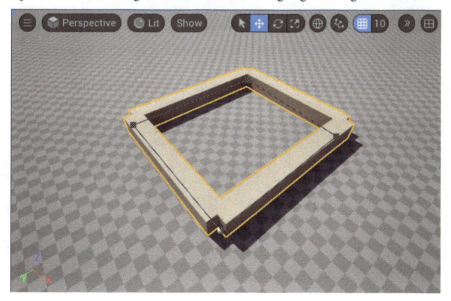

Figure 7.61 – The results after assigning SM_WallCube to the Static Mesh Spawner node

Fixing offset geometry

We still need to fix the corners of this structure. That means we have to make a fix for the offset geometry along the spline controller! It's quite a simple fix, so let's head back to the PCG graph and add one more parameter:

1. Click on **Graph Settings** in the top-right corner of your PCG graph, and in the **Instance** tab on the right-hand side panel, add a new parameter. Ensure it is a vector variable and rename it to `FixingPivot`.

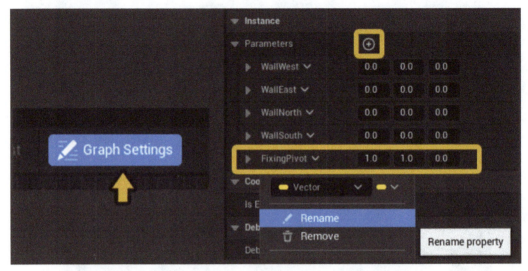

Figure 7.62 – Add the FixingPivot Vector parameter inside the Graph Settings panel

2. The next step involves incorporating the `FixingPivot` variable into your graph. Connect this variable to both the **Offset Min** and **Offset Max** inputs of the **Transform Points** node.

Figure 7.63 – Connect the FixingPivot node to the Transform Points node

3. Let's go back to the `BP_BuildingStructure` actor blueprint and add another **Set Vector Parameter** node inside the **Construction Script** graph! Right-click on the yellow **Value** input pin and choose **Split Struct Pin**. Under the **Name** pin tab, enter `FixingPivot`.

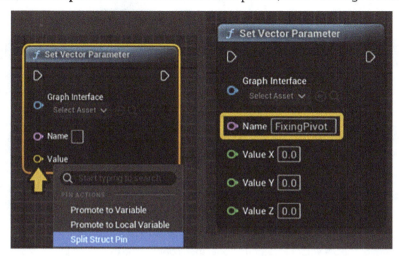

Figure 7.64 – Add the Set Vector Parameter node and the FixingPivot to the Name section

4. Let's use the **Offsetter** macro and connect with the **DimensionOffset** variable. We are going to use the **X** value for **Set Vector Parameter**, so connect the **Offsetter**'s **ReverseDirection** output to the **X** value!

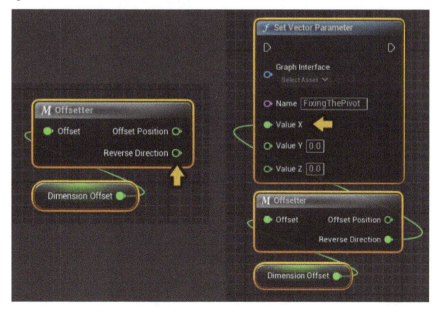

Figure 7.65 – Connect the Offsetter macro's Reverse Direction
output to the X value of Set Vector Parameter

5. Connect the **Graph Instance** node to the Set Vector Parameter node and connect the Set Vector Parameter node to the execution pin.

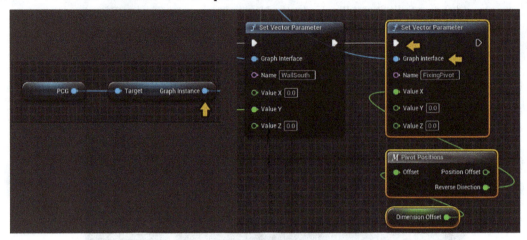

Figure 7.66 – Connect the Graph Instance node to the Graph Instance input
of the Set Vector Parameter node, then connect the Execution pin

Currently, the walls are still misaligned and displaced, leaving gaps at the corners of the building structure. To fix this issue, we will add another instruction to the PCG graph but we will use it toward the end of this chapter. Next, we will add this instruction to our PCG graph and to the BP_ BuildingStructure actor blueprint:

1. In your PCG graph, go to the top-right corner and click on **Graph Settings**. Add one parameter: a vector variable, renaming it to WallOffset. Leave the values set to 0.

Figure 7.67 – Adding a Vector variable and renaming to a WallOffset parameter

2. Now, we can add the **WallOffset** parameter to our graph. We can also add the **Transform Points** node to the graph. Connect **WallOffset** to **Offset Min** and **Offset Max**, respectively.

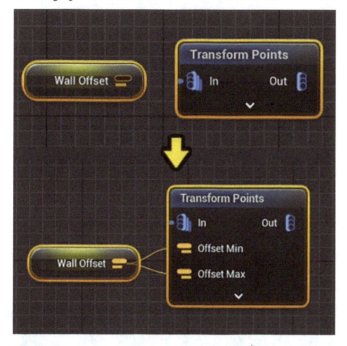

Figure 7.68 – Connecting Wall Offset Vector parameter to the Offset Min and Offset Max input of the Transform Points node

We will connect this node structure to the **Static Mesh Spawner** node later. For now, we can move on to the next section, which will be about **Vertical Expansion**.

We're making great progress here! In the upcoming section of our tutorial, we'll introduce additional nodes to your PCG graph. These nodes will enable us to construct a building structure vertically, based on the number of floors. Stay tuned as I guide you through the process in the next part!

Vertical expansion

In this section, we will go through adding parameters in the **Graph Settings tab** inside your PCG graph, which we will use in your actor blueprint. Before getting into the actor blueprint, let's add a few more nodes to the PCG graph and connect it with the existing graph. Without further ado, let's get started:

1. Inside the PCG graph, go to **Graph Settings**. Let's add two parameters: the **Height** and **CopyTransform** variables. Make sure to set both variables to **Vector**! These variables will be a crucial part of creating new walls, which will be transformed upward!

Figure 7.69 – Add the Height and Copy Transform parameters

2. Open your BP_BuildingStructure actor blueprint and add two **Set Vector Parameter** nodes. Make sure to put the right names for the **Name** field inside the nodes. One should be for **Height** and another for **CopyTransform**.

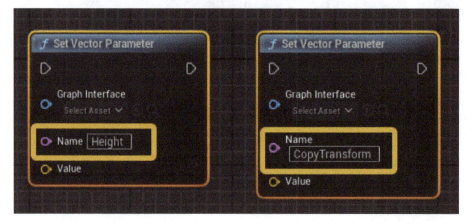

Figure 7.70 – Add the names for the Height and CopyTransform for each Set Vector Parameter node

3. In the **Set Vector Parameter** node, find the section that is responsible for the **Height** variable. Right-click on **Value** and choose the **Split Struct Pin** option. Since the height of the structure is moving vertically, we will only use the **Z** value!

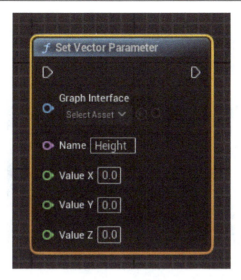

Figure 7.71 – Split struct pin for the Height Set Vector Parameter node

4.   Drag and drop two variables (**Dimension Offset** and **NoOfFloors**) to the graph. Right-click, add a **Multiply** node, and connect them with two other nodes. This will allow us to duplicate and multiply the walls structure upward!

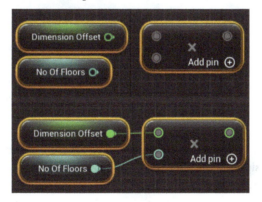

Figure 7.72 – Add the Dimension Offset float variable and the No Of Floors integer
variable nodes. Then, connect these nodes to the inputs of the Multiply node

5. With that, let's connect those nodes to the **Z** value of the **Set Vector Parameter** node, just as is shown in *Figure 7.73*!

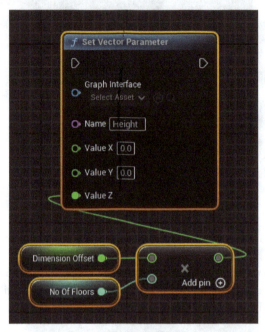

Figure 7.73 – Connect the newly created calculus node structure to the
Value Z input of the Height Set Vector Parameter node

6. To fix walls drifting apart when moving up, add the **GetActorLocation** node and **Multiply operator** to the **Set Vector Parameter** node with the **CopyTransform name**. Right-click on the second input pin in the **Multiply** node and change it to the **To Float(single-precision)** input value. This tweak will keep the walls in line when you're playing around with multiple floors in the BP_BuildingStructure actor blueprint. Make sure to put the -1.0 value for the float value inside the **Multiply** node. Now, let's connect this node to the **Value** input in the **Set Vector Parameter** node!

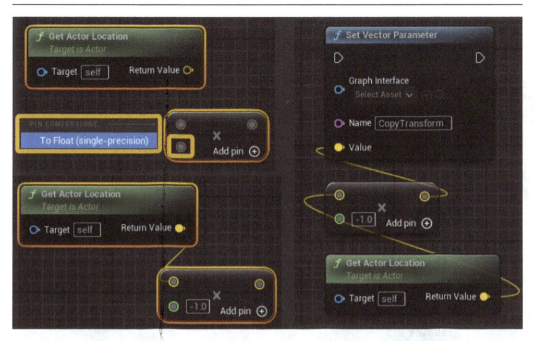

Figure 7.74 – Convert the input node to a ToFloat (single-precision) node and connect it to the Get Actor Location return value, then connect the Multiply node to the Value input

7. Connect the **Set Vector Parameter** node, which uses the **Height** variable, to the **Set Vector Parameters** nodes from the previous examples and to the **PCG Graph Instance** node within the actor blueprint to maintain the flow of the graph. This setup ensures that the **Height** variable is properly integrated and utilized in the graph sequence. Repeat the same process for the **CopyTransform** vector parameter.

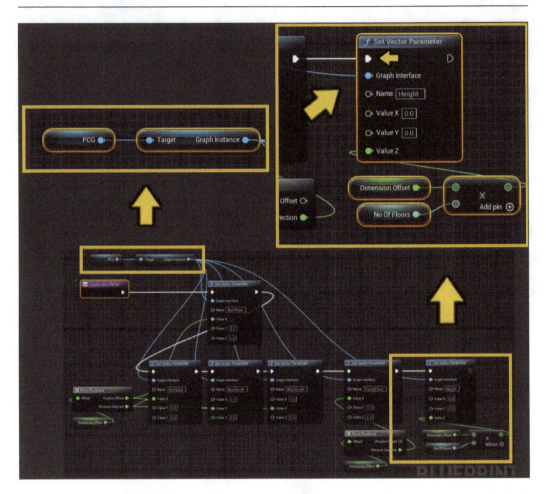

Figure 7.75 – Simplified Graph structure, which represents the right connection
to the graph instance node for the Height Set Vector Parameter node

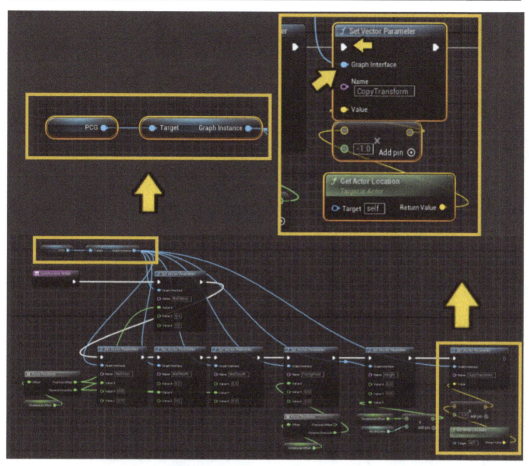

Figure 7.76 – Simplified Graph structure, which represents the right connection to the graph instance node but for the CopyTransform Set Vector Parameter node

8. Let's add another **Set Vector Parameter** node. Right-click on the **Value** input and splits its struct to float variables. Inside the **Name** field, enter `WallOffsetPivot` to match the parameter name from the PCG graph.

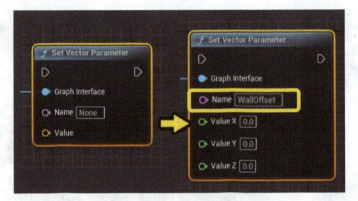

Figure 7.77 – Rename the Name input from None to WallOffset

9. Add two new variables. Call them **HeightGap** and **WallOffset** and make them public. The **HeightGap** variable will be used to fill up any gaps between floor vertical expansion.

Figure 7.78 – Add two float variables and replace their values

10. Drag out our **Offsetter** macro function and connect the **WallOffset** variable to the **Offset** input. Then connect the **Offset Position** output to the **Value X** input of the **Set Vector Parameter** node.

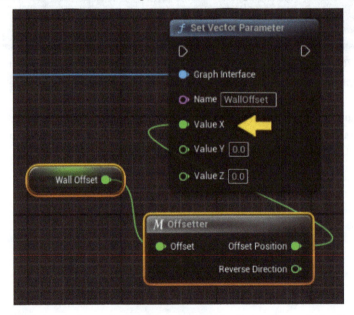

Figure 7.79 – Connect the Offsetter's Offset Position output to Value X

11. The final step is to connect the **Set Vector Parameter** node to the rest of the continuous chain, right after the **Set Vector Parameter** node for the **Copy Transform** instruction, as shown in the following figure. Once connected, save and compile the BP_BuildingStructure blueprint.

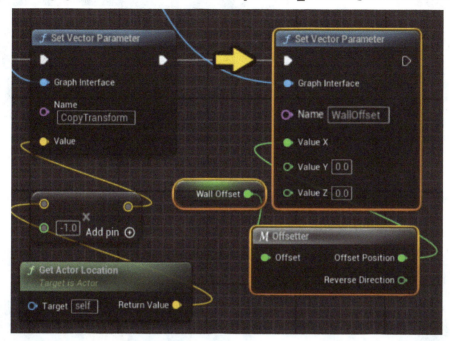

Figure 7.80 – Connecting the Set Vector Parameter nodes together to form a continuous chain

Duplicating walls

We'll resume our work in the PCG_BuildingStructure graph to continue setting it up. It will let us concentrate on developing other parts of our graph.

This time, our primary focus will be on duplicating the walls vertically based on the number of floors, which we can accomplish using the PCG graph. Let's proceed with our work on the PCG graph to implement this feature:

1. In your PCG graph, add the following nodes: **GetActorData**, **MergePoints**, and two **TransformPoints** nodes.

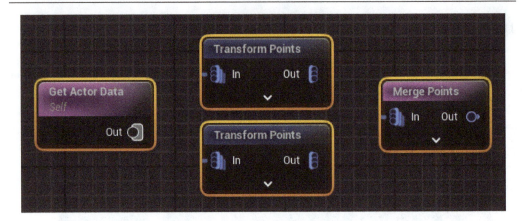

Figure 7.81 Adding the nodes inside the PCG graph

2.   Select the **Get Actor Data** node, and on the right-hand side under **Data Retrieval Settings**, change **Mode** to **Get Single Point**. This will help to produce a single point per actor with the actor transform and bounds.

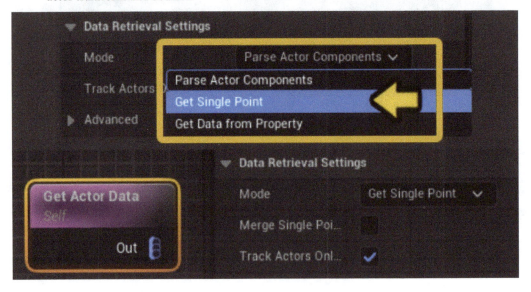

Figure 7.82 – Change the mode to Get Single Point in the Get Actor Data node

3.   Connect all the nodes to form a mini graph. Let's add a **Height** vector parameter to this graph and connect it to the bottom **Transform Points** node's **OffsetMin** and **OffsetMax** input.

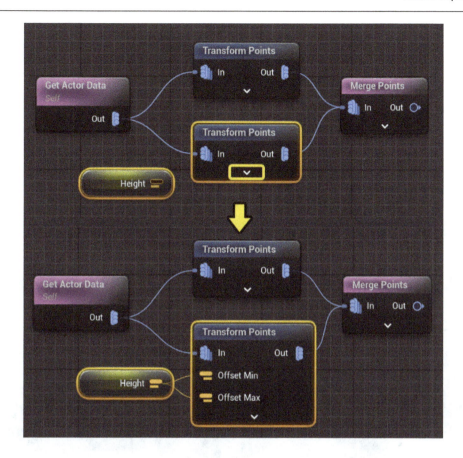

Figure 7.83 Connect the Height Property node to both the Offset Min and
Offset Max input properties inside the Transform Points node

4. Let's introduce a few more nodes into our graph. Add the **Create Spline** and **Spline Sampler** nodes to the graph.

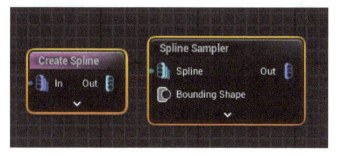

Figure 7.84 – Add the Create Spline and Spline Sampler nodes inside the PCG graph

> **Note**
>
> You may occasionally encounter errors or warning messages on some nodes. Often, this is due to a glitch where Unreal Engine struggles to refresh them properly. Typically, these issues don't cause any harm to the graph and can be resolved by manually refreshing the nodes or restarting the editor if necessary.

5. Select the **Spline Sampler** node. In the **Settings** panel, change **Dimension** to **On Spline**. In the **Mode** settings, change it to **Distance** and set its **Distance Increment** value to 200 units. `This will represent a gap distance between the new walls created vertically on each floor. Make sure to activate the Unbounded setting

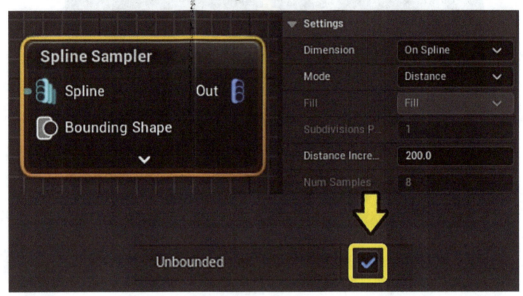

Figure 7.85 – Set Dimension to On Spline and Mode to Distance

6. Connect all the nodes in the same manner as shown in *Figure 7.86*!

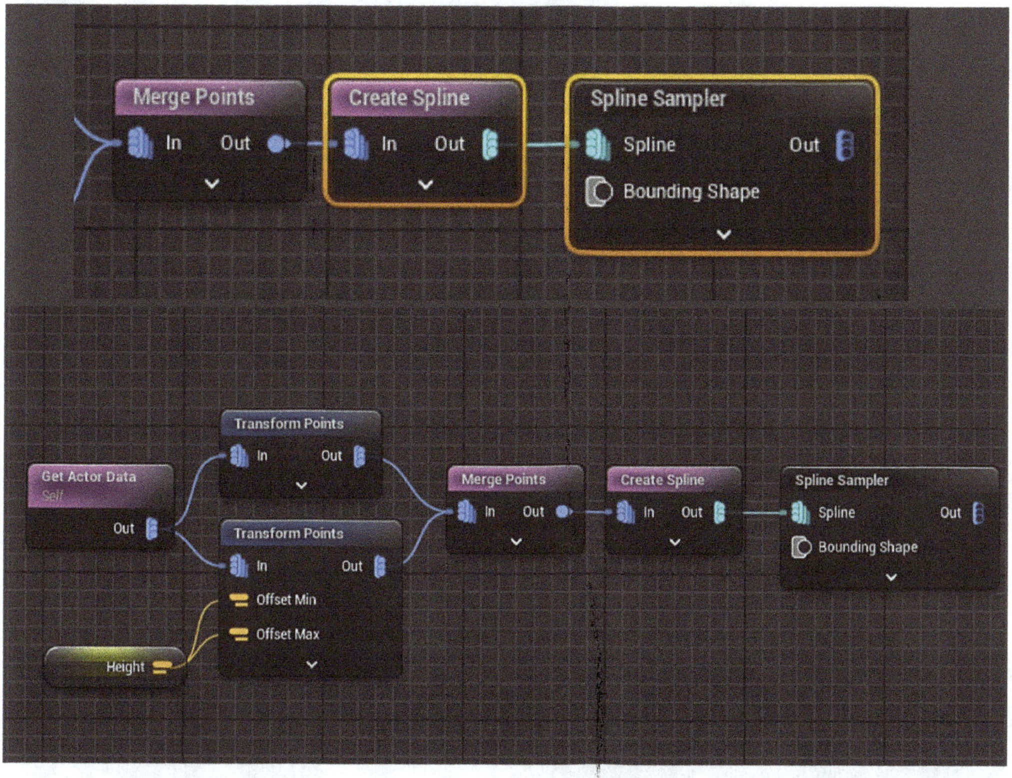

Figure 7.86 – Connect Create Spline and Spline Sampler with the Merge Points node

7. To finalize this part, let's add one more parameter to our PCG graph. In our PCG graph, right-click and add a **Get Actor Property** node. On the right-hand side, under **Settings**, change **Property Name** to **HeightGap** to match the variable name we created inside our BP_BuildingStructure blueprint.

Figure 7.87 – Adding Get Actor Property and renaming its Property Name name to HeightGap

Connect your **GetActorProperty** node to the Spline Sampler's **Distance Increment** input

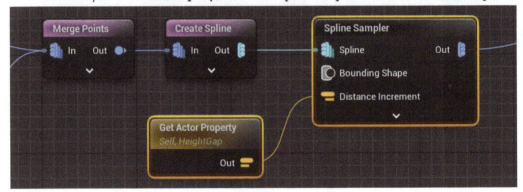

Figure 7.88 – Connecting Get Actor Property to the Distance Increment output of the Spline Sampler node

This was our final setup to ensure that our building structure can offset the height gap position while also giving us more control over the height differences when new static meshes are spawned on the building. In the next part, we'll go through the steps of combining the walls and bringing the building structure to life!

Combining wall meshes

In the next phase of this exercise, we will focus on copying the height points and combining the wall meshes. This will enable us to spawn the walls vertically according to the number of floors specified. We'll achieve this by using the height data to determine the vertical duplication of wall meshes, effectively stacking them to form the floors of the building structure in our PCG graph:

1. Add the **Copy Points** and **Transform Points** nodes, as well as the **Copy Transform** parameter, to your graph. The **Copy Transform** parameter will play a major role in fixing the positions of the walls being spawned upward.

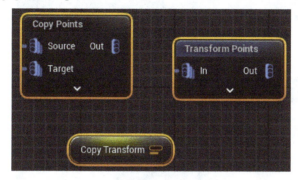

Figure 7.89 – Add Copy Points and Transform Points, as well as the
Copy Transform property node, inside the PCG graph

2. Select the **Copy Points** node. On the right-hand side in the **Settings** panel, change **Rotation Inheritance** to **Target**.

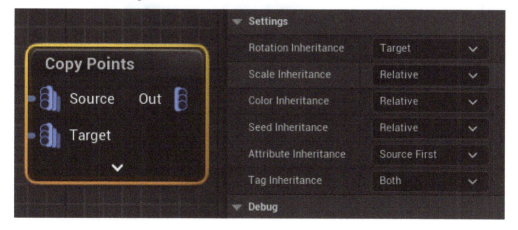

Figure 7.90 – Change the Rotation Inheritance setting of the Copy Points node to Target

3. Connect the **Copy Transform** parameter to the **Transform Points** node's **OffsetMin** and **OffsetMax** inputs.

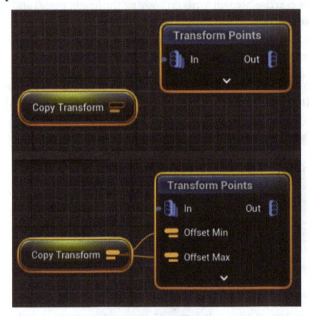

Figure 7.91 – Connect the CopyTransform property node to both the Offset Min and Offset Max inputs in the Transform Points node

With all this in place, connect the **Copy Points** node to the **Transform Points** node.

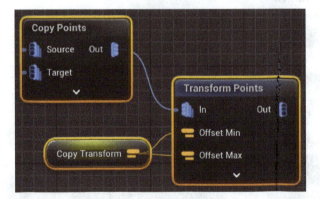

Figure 7.92 – Connect the Copy Points output to the Transform Points input

4. Let's introduce and add a new node—a **Switch** node! The **Switch** node functions as a control flow mechanism. It passes input data to a specific output based on a selection that matches the corresponding property. This node is essential for dynamically directing the flow of data within the graph based on variable conditions or criteria.

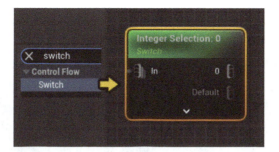

Figure 7.93 – Add the Switch node inside the PCG graph

5.   In the section where we previously connected the **Static Mesh Spawner** node to **Transform Points** with the **Fixing Pivot** parameter, let's disconnect the **Static Mesh Spawner** node and add the **Switch** node to that part instead. The **Switch** node will help in managing complex PCG graphs by reducing the need for multiple branches and redundant nodes.

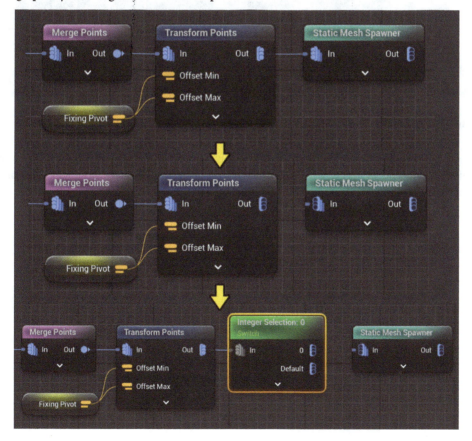

Figure 7.94 – Disconnect the Static Mesh Spawner node from the Transform Points node and connect the Switch node to the output of the Transform Points node

6. Bring the **Copy Points** node section closer to the **Switch** node. Connect the 0 output of the Switch node to the Target Input of the Copy Points node.

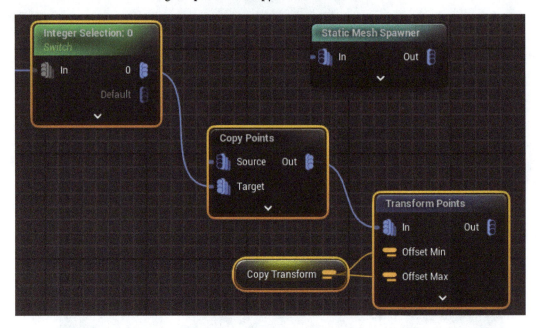

Figure 7.95 – Connect the output of the Switch node to the Target input of the Copy Points node

7. Go back to your section of the graph mentioned in *Figure 7.86* that you have created for setting up the height and its structure vertically and connect its **Spline Sampler** output with the **Source** input of the **Copy Points** node.

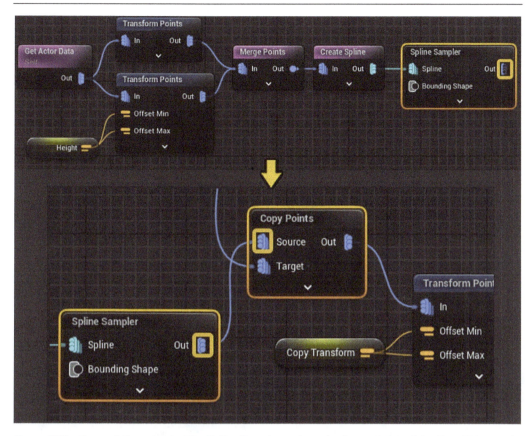

Figure 7.96 – Connect the output of the Spline Sampler node to the Source input of the Copy Points node

Here's a summary of all the connected nodes you should have by the end of this step. We are duplicating the points vertically according to the number of floors, aligning them with the specified height values from the Spline Sampler's distance increment.

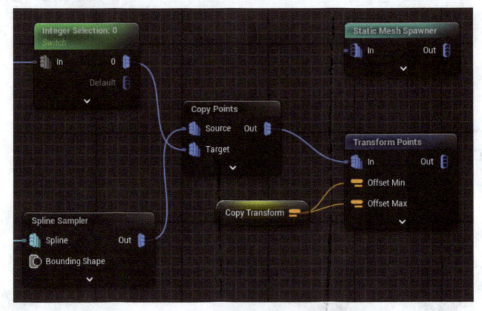

Figure 7.97 – Final layout of the Copy Points structure within the PCG graph

Wrapping up the structure

In the next part, we are going to finish the entire building structure exercise by adding an extra few nodes that will help to generate the window gaps between the walls. In this case, we are going to pair an entire graph to form a streamlined connection with the flow of the entire graph and connect the **Static Mesh Spawner** node back to the graph connection:

1.   Let's add two nodes to our **Difference** graph and the **Density Filter** node.

Figure 7.98 – Add the Difference and Density Filter nodes to the PCG graph

2. From the Content Browser under the **PCG | Blueprint** folder, use the **SetToPoints** blueprint. Drag and drop it to the PCG graph space. Make sure to place it next to the nodes from the previous example.

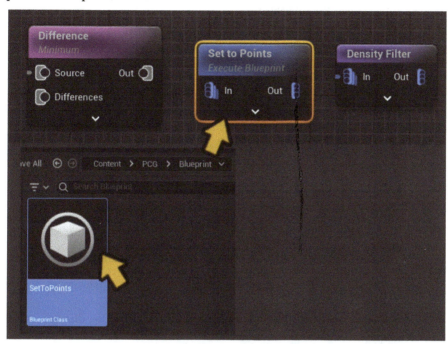

Figure 7.99 – Add the SetToPoints PCG blueprint to the PCG graph
between the Difference and Density Filter nodes

3. Connect the nodes to form a straight-line graph where all the nodes are placed next to each other, as shown in *Figure 7.100*.

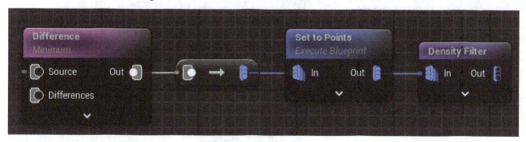

Figure 7.100 – Connect the output of the Difference node to the input of the SetToPoints node,
and then connect the output of the SetToPoints node to the input of the Density Filter node

4. When you use the **SetToPoints** node, you can tweak your window points and density settings to make the gaps around the windows bigger on each wall side. The **Density Filter** node works similarly, allowing you to control how frequently the window gaps appear within the structure's alignment. Set its **Lower Bound** value to 0.2.

Figure 7.101 – Change the settings for the SetToPoints and Density Filter nodes

5. With all this set up, connect the **Static Mesh Spawner** node, which we disconnected in the previous example, to the **Density Filter** output.

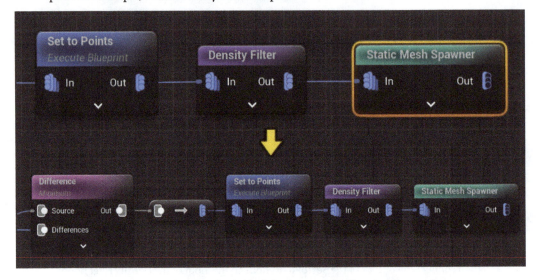

Figure 7.102 – Connect Static Mesh Spawner to the Density Filter node

6. We have also one left instruction node that we have already created in the previous section, under **Fixing Geometry Offset**. Let's connect it between the **Static Mesh Spawner** and **Density Filter** nodes.

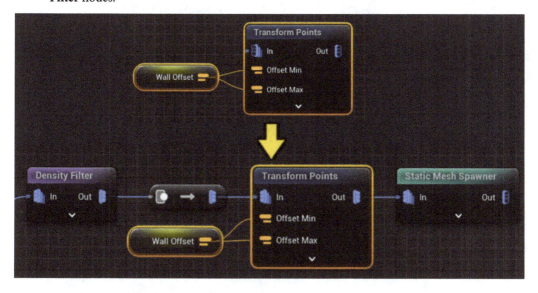

Figure 7.103 – Connecting the previously formed WallOffset Transform Points node between the Density Filter and Static Mesh Spawner nodes

We've successfully aligned all the nodes in a single line, and now we'll connect all the earlier graph nodes to the inputs for **Source** and **Differences** on the **Difference** node. This step is key—it'll turn our setup into a functional PCG graph that syncs up with our BP_BuildingStructure actor blueprint. Let's dive in and get this done:

1. Connect the **Switch** node's **Default** output to the **Differences** input on the **Difference** node. Then, link the output of the **Transform Points** node to the **Source** input of the **Difference** node. This will ensure that your node transformations are correctly applied.

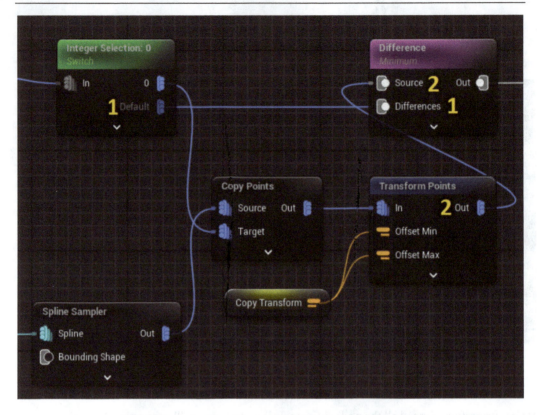

Figure 7.104 – Connect the outputs of the Transform Points and
Switch nodes to the inputs of the Difference node

2. Let's test out your BP_BuildingStructure actor blueprint on the scene and check out the
    results. Also, change your values for the number of floors in your **Details** panel and observe
    the changes.

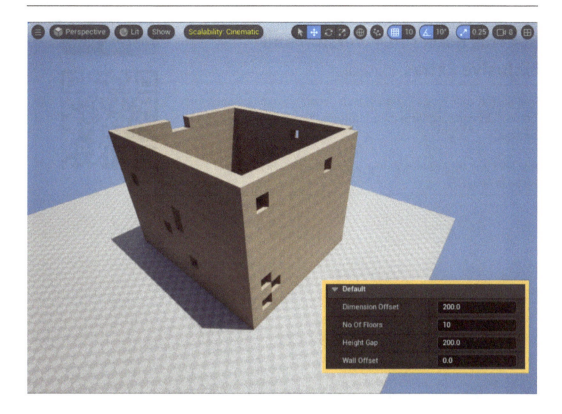

Figure 7.105 – Testing the building structure actor blueprint in the scene

You've just wrapped up this chapter, and you've made some impressive progress through this rather challenging exercise! It's great to have this knowledge in your toolkit, especially since you'll use this example as a template for building more complex structures. You've definitely picked up some valuable lessons along the way!

Summary

In this chapter, you picked up essential techniques for using spline control to manage the PCG graph. You've learned how to configure your actor blueprint to construct building structures effectively using the PCG graph. This approach is particularly powerful when creating numerous building structures across the landscape, allowing you to adjust asset quantities from the viewport in real time as you build the city.

In the next chapter, we'll explore how to create a PCG biome using custom PCG graphs that integrate seamlessly with textures specifically designed for landscapes in Unreal Engine 5!

Get This Book's PDF Version and Exclusive Extras

UNLOCK NOW

Scan the QR code (or go to `packtpub.com/unlock`). Search for this book by name, confirm the edition, and then follow the steps on the page.

Note: Keep your invoice handy. Purchases made directly from Packt don't require an invoice.

8

Building Biomes: Mastering PCG for Rich Environments

Welcome to *Chapter 8*! After mastering structure building with spline controllers in *Chapter 7*, we're ready to take on a new challenge. In this chapter, you'll learn how to create diverse and dynamic biomes using the PCG Biome plugin, perfect for those eager to design rich and varied landscapes with procedural generation techniques.

We'll explore generating various biome environments by using datasets and texture information to shape terrain variations and foliage distribution. You'll develop methods to control and customize these elements, creating unique and immersive biomes without relying solely on the PCG graph.

This chapter is divided into sections that cover foliage placement based on texture information. You'll get hands-on experience with new PCG techniques, learning how to manage PCG methods that bring assets together into cohesive and realistic biomes.

Additionally, all the chapter's content will be available via the GitHub link at the end of the tutorial. Get ready to dive into the exciting world of biome creation, enhancing your landscape designs with the PCG Biome plugin and texture information!

Without any further ado, let's get started!

This chapter concentrates on the following key areas:

- Getting familiar with the foliage assets
- Understanding the PCG Biome plugin
- Creating PCG Biome blueprints and your own PCG Biome using data assets
- Generating biomes based on texture landscape information

By the end of this chapter, you'll gain new techniques and valuable insights into setting up your own PCG Biome and using data assets to control the flow of the foliage.

Technical requirements

You will need the following to complete this chapter:

A good computer that can run PCG and those are with a a multi-core CPU (AMD Ryzen 7/9, Intel i7/i9), GPU (NVIDIA RTX, AMD Radeon RX with 8+GB VRAM), and at least 16 GB RAM

- A basic understanding of setting up blueprints in Unreal Engine 5

- The template project includes modified models from the PCG Plugin folder. Additionally, you can download your favorite assets from Megascans if you want a variety of foliage in your scene.

The Unreal Engine version that's used in this chapter is version UE 5.4. PCG Biome was introduced with this version of Unreal, hence the latest version of Unreal Engine is the most beneficial for this chapter.

For this project, we will use the template project, which is available in the GitHub repository, and you can download its template from the following GitHub link:

`https://github.com/PacktPublishing/Procedural-Content-Generation-with-Unreal-Engine-5/tree/main/Chapter_8/UE5_PCG_Chapter_08`

The code in action video for the chapter can be found at `https://packt.link/fiOoY`

Getting familiar with the foliage assets

As we progress through this chapter, we'll be using PCG model assets that I have specifically prepared for this exercise. These models originate from the PCGPplugin folder, but I have modified them to better suit our needs:

1. Navigate to the `Models` folder and you will find two models, which are called `PCG_Bush_01` and `PCG_Tree_02`.

Figure 8.1 – Getting familiarized with the PCG models

2. As part of the model preparation process, we have to enable **Nanite** mode for each provided mesh in this example. Like we always do in this part of the exercise, right-click on each mesh and enable **Nanite** mode.

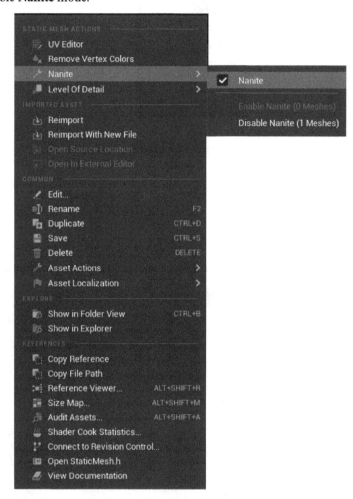

Figure 8.2 – Enable Nanite mode on each static mesh

LandscapeTexture

This exercise focuses on distributing foliage spawn across the landscape using a specific approach. What's interesting is that the plugin allows projecting texture information onto the landscape, enabling it to generate models based on the color assignments in the texture. Each color in the texture corresponds to specific model information, guiding the placement of the models. For this exercise, we will use a simple texture that I quickly created in Photoshop.

Figure 8.3 – Landscape texture for the Biome PCG exercise

In the next section, I will walk through and explain which color will correspond to different models in this project.

Project landscape overview

The project comes with the ready-made landscape terrain and it has the landscape material assigned to it.

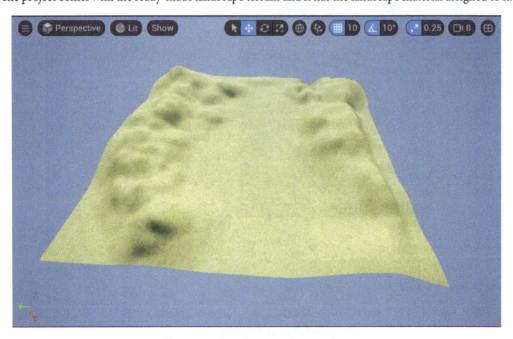

Figure 8.4 – Landscape actor overview

In the following section, we will set up the PCG Biome plugin and configure the texture using the PCG graph, which is specifically designed to integrate with the landscape texture.

Setting up the PCG Biome plugin

The PCG Biome plugin is a separate extension that comes with the latest version of Unreal Engine 5.4 and it helps users to create a foliage distribution across the landscape much faster! For this reason, let's enable this plugin in our project:

1. At the top of your window, go to **Edit|Plugins**. In the **Plugins** window, search for the **PCG Biome Core** plugin. It will ask you to restart your project, so click **Restart**!

Figure 8.5 – Enable the PCG Biome Core plugin

2. We will use one of the PCG assets to work with this example, which we will find inside the Biome plugin folder. First, let's reveal the Plugin Content and Engine Content inside the project. Click on the **Settings** button and enable the Plugin Content and Engine Content folders.

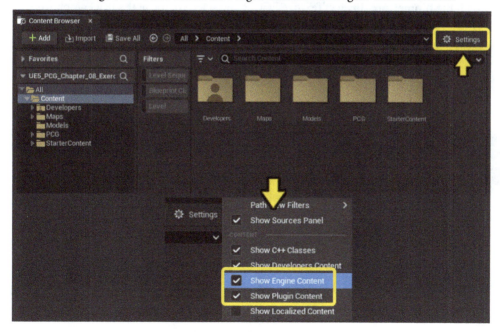

Figure 8.6 – Enable the Show Engine Content and Show Plugin Content options

3. The next part is to search for the PCG graph that we want to use for this exercise. Navigate to the **Engine -> Plugins -> PCG Biome Core Content -> Core** folder and search for the **BiomeCore_ProjectTexture** PCG graph.

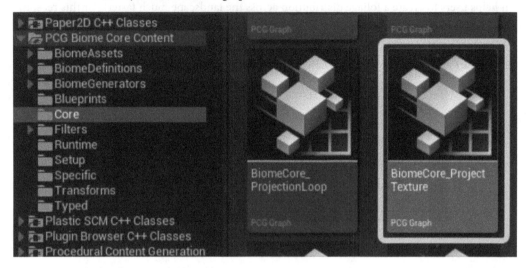

Figure 8.7 – Locating BiomeCore_ProjectTexture

4. With the **BiomeCore_ProjectTexture** PCG graph selected, move the PCG graph from the **Plugins** folder to the **PCG | TextureBiome** folder.

Figure 8.8 – Migrating the BiomeCore_ProjectTexture PCG graph
from the Plugins folder to the TextureBiome folder

With the PCG graph from the **Plugins** folder sorted, now we are ready to dive into this chapter. In this section, we'll walk through the project and begin setting up the data assets and blueprints included. We'll start by applying a texture to the landscape, which will project models onto it. I will guide you through each step of this process. So, let's get started!

Biome Core

Before we can start working with the PCG Biome plugin, we have to first create the blueprint actor that consists of the PCG component responsible for generating the biome. To do this, we must add it to our Content folder:

1. Inside the **Content | PCG** folder, right-click and select the blueprint class. On the blueprint class menu, select the search bar, search for BP_PCGBiomeCore, and add it to your PCG folder:

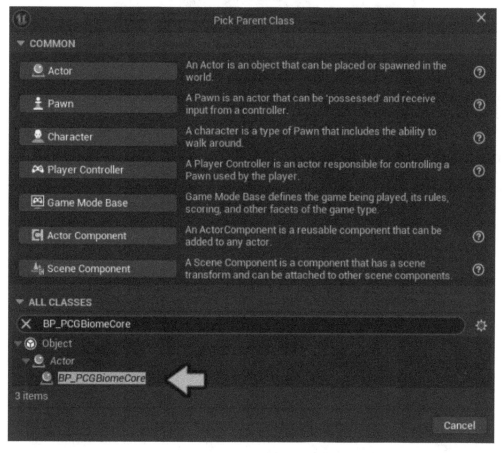

Figure 8.9 – Add the BP_PCGBiomeCore blueprint

2. Select the newly created blueprint and rename it to a similar name, such as `BP_BiomeCore`. For now, drag and drop the **BP_BiomeCore** blueprint into the scene and save the scene.

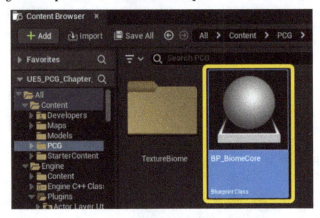

Figure 8.10 – Drag and drop the BP_BiomeCore actor blueprint into the scene

We have successfully created the **BP_BiomeCore** blueprint and placed it in the scene. The next part is to create a PCG Instance graph, which we will use to project our texture sample – the one that I have provided with the project.

Biome Texture projection

This part of the PCG realm is pretty cool because it lets you procedurally generate foliage and other meshes on the landscape just by assigning texture information to the terrain! As you might have noticed, this is similar to what we did in *Chapter 6*. This time, we'll be using a pre-made PCG graph from the PCG Biome plugin templates that we have moved from the Biome plugins folder. So, without further ado, let's get started!

1. Inside your **PCG | TextureBiome** folder, right-click with your mouse button, go to the **PCG** tab, and add **PCG Graph Instance**. Rename it to `PCG_TextureBiome`.

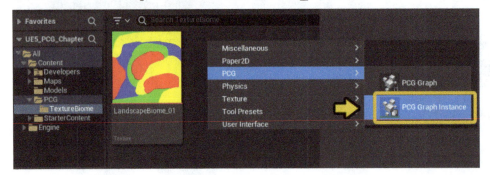

Figure 8.11 – Add PCG Graph Instance

2. Double-click on your newly created PCG Graph Instance and let's add the **BiomeCore_ ProjectTexture** PCG graph from the TextureBiome folder.

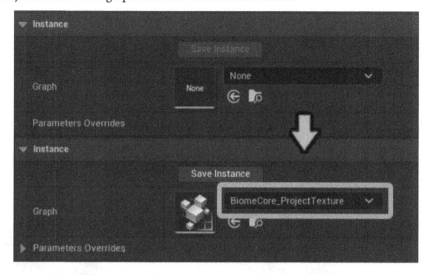

Figure 8.12 – Assign the BiomeCore_ProjectTexture PCG Graph
inside the PCG_TextureBiome Graph Instance

3. With this all set up, click on the **Parameters Overrides** panel to scroll down all the options available for this PCG Graph Instance configuration.

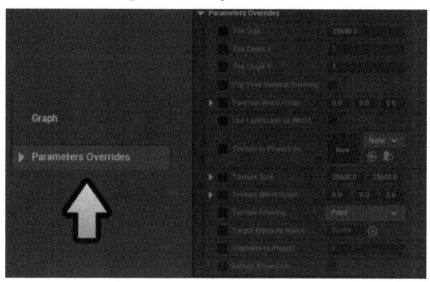

Figure 8.13 – Expand Parameters Overrides

4. To start, the texture size should match the landscape's area. In other words, the texture should be proportional to the landscape – for example, if the landscape resolution is 505 by 505 units, the texture size of the current PCG

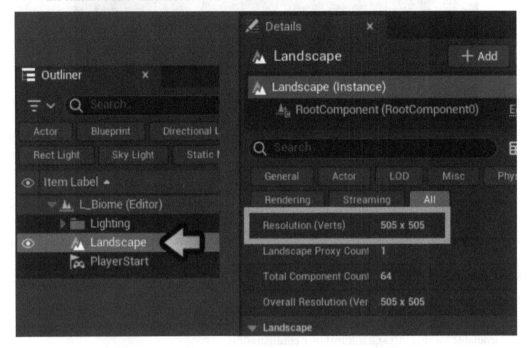

Figure 8.14 – Check the Resolution (Verts) values for the Landscape actor

Graph instance should be double the size. You can check this by selecting the landscape in the outliner and looking at **Resolution (Verts)** *(Figure 8.14)* in the **Details** panel.

Figure 8.15 – Check Texture Size

5. The next step is to activate the **Texture to Project On** and **Texture Size** parameters. Let's add the **LandscapeBiome_01** texture to the available slot. Then, crank the value of **Texture Size** up to 51200.0 and 51200.0, respectively.

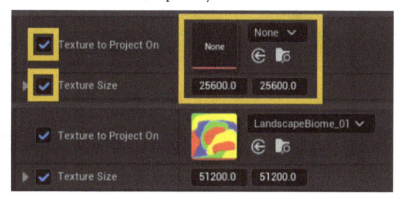

Figure 8.16 – Enable the Texture to Project On and Texture Size options. Change Texture Size to 51200.0

6. Now that we've set the important parameters, we need to enable a few more settings to make the texture work with the landscape. Let's turn on **Use Landscape as World Origins** and **Channels to Project**. These settings help arrange the texture information to properly align with the landscape terrain geometry in three-dimensional space.

Figure 8.17 – Enable the Use Landscape as World Origins and Channels to Project settings

7. As you may have noticed, **Channels to Project** is set to the x value. That means it will only work on the x axis. To make sure it works on all axes, we have to add the y and the z values next to the already existing x!

Figure 8.18 – Change Channels to Project to the xyz values

Now you can save the PCG_TextureBiome graph instance, and we can move on to enabling the graph instance in the **BP_BiomeCore** blueprint. In the next section, I will illustrate this step by step and show you how to achieve this.

Biome texture setup

In this section, we'll explore a cool feature: texture projection. I'll show you how to enable it and debug the visuals of the landscape texture formation. Let's grab our **BP_BiomeCore** blueprint from the Outliner and enable a few settings:

1. Select the **BP_BiomeCore** blueprint in the Outliner. In the **Details** panel, under the **BP_BiomeCore (Self)** actor, search and select the **BiomeCore** PCG component.

Figure 8.19 Select BiomeCore on the Details panel of the BP_BiomeCore blueprint

2. With the **BiomeCore** PCG component selected, navigate down to the **Instance** tab and open the **Parameters Overrides** tab. Enable **Biome Texture Projection Instance**. For this, we will choose the **PCG_TextureBiome** graph, which we created in the previous section.

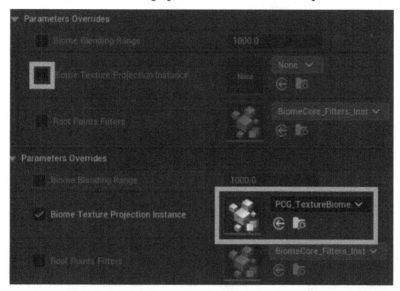

Figure 8.20 – Enable Biome Texture Projection Instance and add the PCG_TextureBiome graph

3. To test our first BiomeCore blueprint, it will be better to enable the debug mode. In the same **Instance** tab location, search for **Debug- Display Biome Cache** and enable it just like in the following screenshot.

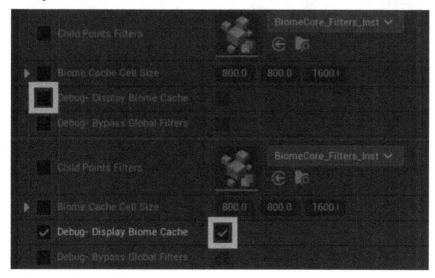

Figure 8.21 – Turn on Debug- Display Biome Cache

4. In the **Details** panel, scroll up to the **PCG** section and click the **Generate** button to produce the desired results.

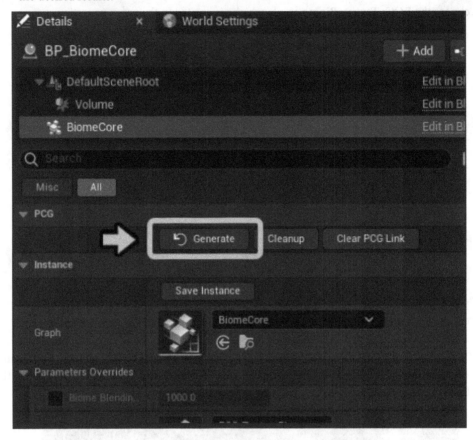

Figure 8.22 – The Generate button is found under the PCG section
of the BiomeCore component in the Details panel

You can now check the results and see for yourself how the texture is used to spawn specific areas based on each color factor. This will be a relevant step in the next part of the integration, where you will learn how to define each color as a separate area to spawn different meshes.

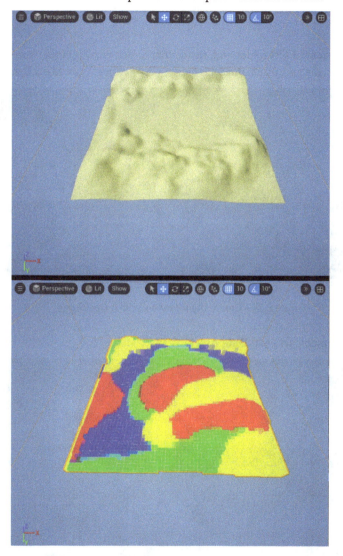

Figure 8.23 – Preview of the landscape texture via Debug mode

You just completed the texture project Biome and you have learned how to use this form of PCG configuration in the landscape terrain. In the upcoming section, you will learn how to use data assets and the Biome setup blueprints to spawn different meshes on the landscape based on coloration tags. Let's get started.

Setting up the Biome Tree

In this section, we are going to explore how to add the meshes to our biome setup. In this case, we need to divide it into other biomes that are assigned to each mesh. In this case, we are going to start with the trees:

1. Go to the **Content | PCG** folder and, inside that folder, create another subfolder. Name this folder TreeBiome and, inside this folder, right-click and search for **Miscellaneous|Data Asset**.

Figure 8.24 – Create the folder and call it TreeBiome

2. On the search menu, let's add **Data Asset** instance.

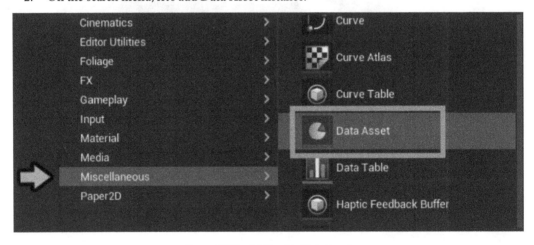

Figure 8.25 – Adding for DataAsset

3. This will open another selection window with a different type of class for each different dataset. For this one, we will choose **BiomeAssetTemplate**. This will be used for spawning the tree meshes on the scene.

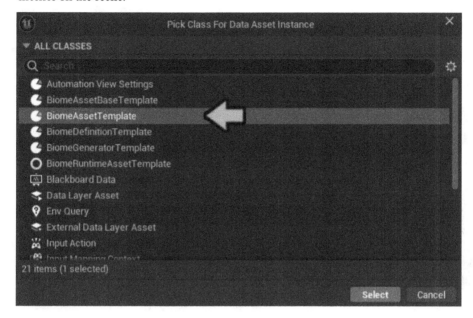

Figure 8.26 – Search for BiomeAssetTemplate

4. Rename your data asset to DA_BiomeATrees. Let's open it and select the + button to add a new array, which is next to **Biome Assets**. It will show the necessary configuration for us to work with.

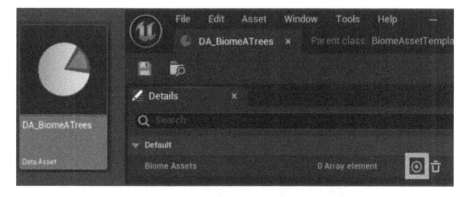

Figure 8.27 – Renaming the DataAsset Instance to DA_BiomeATrees

5. In the **Biome Assets** menu, under the **Generator** slot, add **DefaultGenerator**, which will appear on the drop-down menu. Under the **Mesh** slot, add the **PCG_Tree_02** mesh model.

Figure 8.28 – Assign DefaultGenerator and PCG_Tree_02 to the Generator and Mesh placeholders

6. Let's scroll down and open the **AssetOptions** menu. In this section, we would like to select the **SelfPrune** option, which will remove overlapping assets that will come into contact with the assets from the outer biome setup.

Figure 8.29 – Access AssetOptions and enable the SelfPrune functionality

There is not much we have to do in this part so we may as well close our data asset and carry on with the next part! We will create another data asset, but this time we will use it to define the asset that will be assigned to the spawning trees by using the colors from the landscape texture:

1. Back in the `TreeBiome` folder, right-click and select **Miscellaneous->Data Asset**.

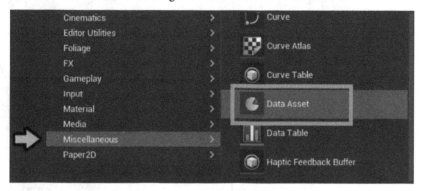

Figure 8.30 – Under the Miscellaneous category, select Data Asset

2. From the drop-down menu of the **Data Asset Instance**, select the **BiomeDefinitionTemplate** class.

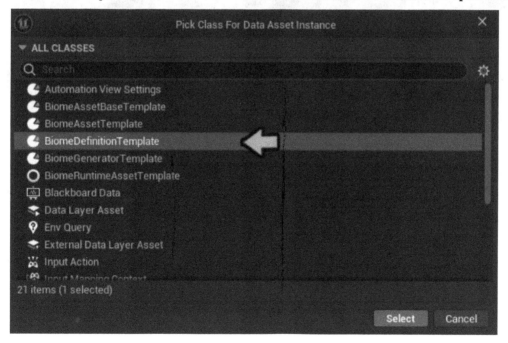

Figure 8.31 – Pick BiomeDefinitionTemplate for the data asset

3. Rename your data asset to DA_BiomeDTrees. Open it and click the arrow button on the left of **Biome Definition**. This will open a simple setup configuration for the landscape texture based on color assignment.

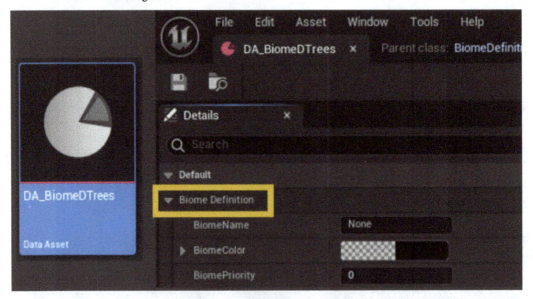

Figure 8.32 – Access Biome Definition inside the DA_BiomeDTrees asset

4. Let's rename **BiomeName** to Trees and change **BiomeColor** to a bright green color (**R:** 0.0, **G:** 1.0, **B:** 0.0, **A:** 1.0), as shown in the following screenshot.

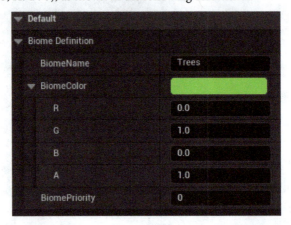

Figure 8.33 – Change BiomeColor to a light green color – R:0.0 G:0.0 B:0.0 A:1.0

We just finished setting up the **DA_BiomeDTrees** data asset and now we can carry on and start setting up the PCG biome setup blueprint, but this time only to form the trees on the landscape! Let's get started:

1. Let's go back to the TreeBiome folder, right-click on it, and from the menu, select **Blueprint Class**. Search for BP_PCGBiomeSetup and select it! Rename your blueprint to BP_TreesBiome.

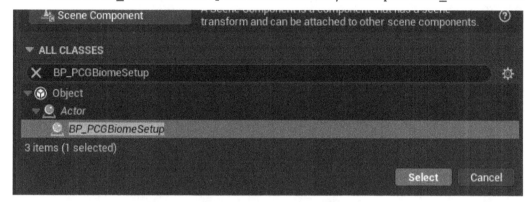

Figure 8.34 – Search for and add the BP_PCGBiomeSetup actor blueprint

Now we have set up the PCG Biome blueprint as **BP_TreesBiome**.

Figure 8.35 – Rename your actor blueprint to BP_TreesBiome

2.   The first thing you will notice is that it's an actor blueprint. The only major difference is that it is programmed to only work with the Data Assets specifically for the PCG use.

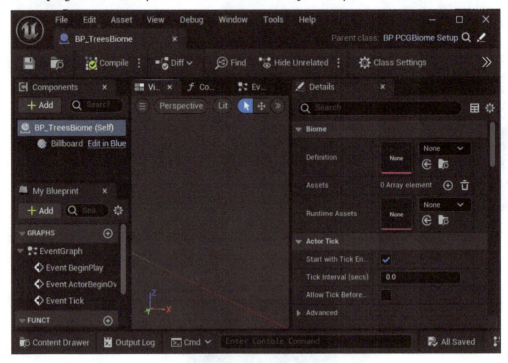

Figure 8.36 – Open the BP_TreesBiome actor blueprint

3.   On the **Details** panel to your left, choose DA_BiomeDTrees for Definition.

Figure 8.37 – In the Details panel, navigate to the Definition tab
and search for the DA_BiomeTrees data asset

4. Just below **Definition**, in the **Assets** tab, add an array element, which will be used for the data assets.

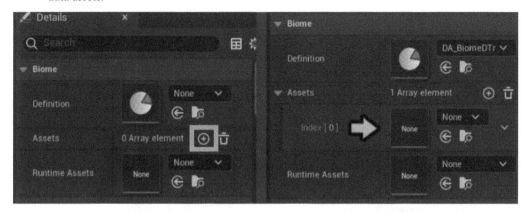

Figure 8.38 – Add an Index [0] array for the new data asset

5. For this index array, let's choose the **DA_BiomeATrees** asset.

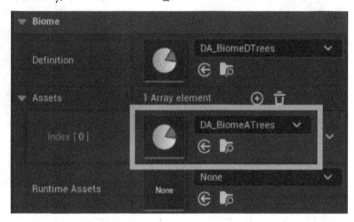

Figure 8.39 – Assign the DA_BiomeATrees asset to the array

6. Let's compile the actor blueprint and return to the main viewport! Now let's drag and drop **BP_TreesBiome** into the scene. Then, you should be able to see the trees start to appear in your scene – just on the green areas while you are in Debug mode. Very cool, right?

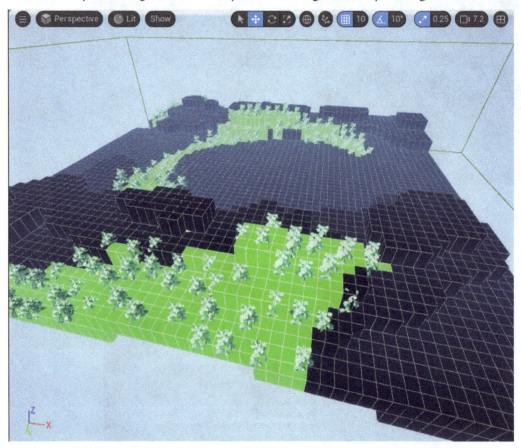

Figure 8.40 – Preview the results of tree spawning in the green sections

You have just completed your first tree biome setup, which will be very helpful for the other parts of the exercise in the next section. We should keep the Debug mode enabled for now to observe how other biomes will behave. Let's repeat the same process for the final setup.

Setting up the Biome Bushes

We are going to use the same process as we used in the previous section, but in this case, we will use bush models instead of the tree models! Let's get started:

1. Go to the **Content | PCG** folder and, inside that folder, create another subfolder. Name this folder `BushBiome` and, inside this folder, right-click and search for **Miscellaneous->Data Asset**.

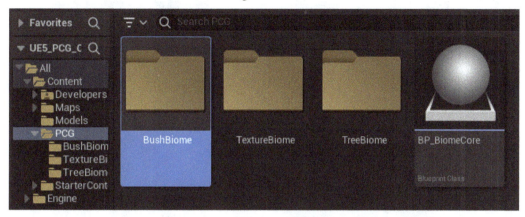

Figure 8.41 – Create a new subfolder and name it BushBiome

2. Next select the **Data Asset** Instance under **Miscellaneous** category

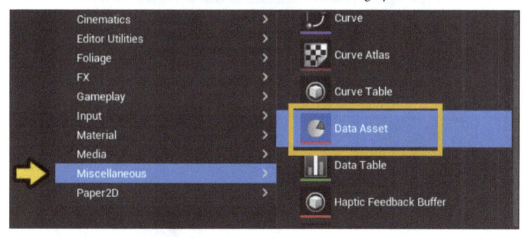

Figure 8.42 – Under the Miscellaneous category, add a new data asset

3. This will open another selection window with different types of classes for each different dataset. For this one, we will choose **BiomeAssetTemplate**. This will be used for spawning the bush meshes in the scene.

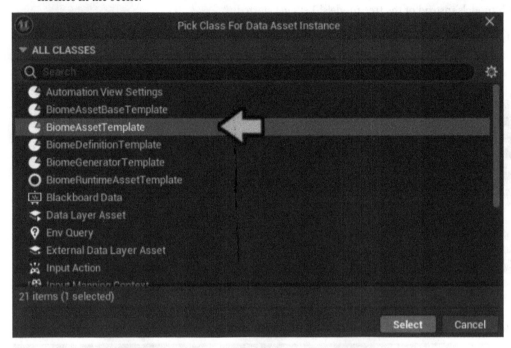

Figure 8.43 – Search for and pick BiomeAssetTemplate

4. Rename your data asset to DA_BiomeABushes. Let's open it and select the + button to add a new array, which is next to **Biome Assets**. It will show the necessary configuration for us to work with.

Figure 8.44 – Inside DA_BiomeABushes, add another array for Biome Assets

5. In the **Biome Assets** menu, under the **Generator** slot, add **DefaultGenerator**, which will appear in the drop-down menu. Under the **Mesh** slot, add the **PCG_Bush_01** mesh model.

Figure 8.45 – Assign DefaultGenerator and PCG_Bush_01 to the Generator and Mesh placeholders

6. Let's scroll down and open the **AssetOptions** menu. In this section, we would like to select the **SelfPrune** option, which will remove overlapping assets that will come in contact with the assets from the outer biome setup.

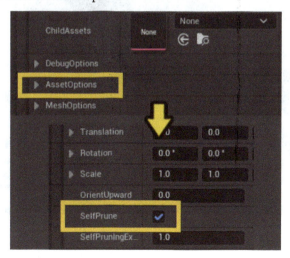

Figure 8.46 – Under AssetOptions, turn on the SelfPrune functionality

There is not much we have to do in this part, so we may as well close our data asset and carry on with the next part! We will create another data asset, but this time we will use it to define the asset that will be assigned to the spawning bushes by using the colors from the landscape texture.

1.   Back in the `BushBiome` folder, right-click and select **Miscellaneous->DataAsset**.

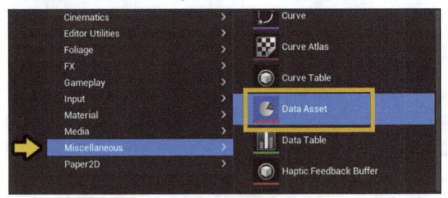

Figure 8.47 – Under the Miscellaneous category, add a new Data Asset

2.   From the drop-down menu of the **Data Asset Instance**, select the **BiomeDefinitionTemplate** class.

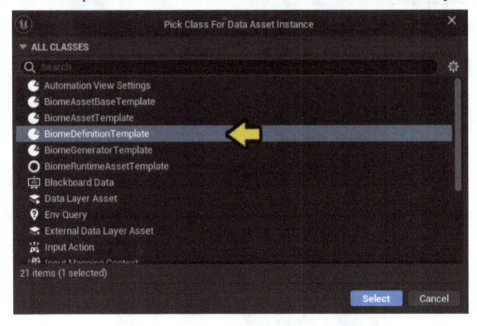

Figure 8.48 – Search for and pick BiomeDefinitionTemplate

3. Rename your data asset to DA_BiomeDBushes. Open it and click the arrow button on the left of **Biome Definition**. This will open a simple setup configuration for the landscape texture based on color assignment.

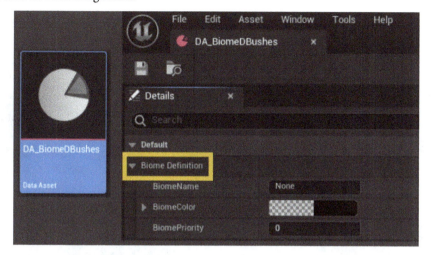

Figure 8.49 – In the DA_BiomeDBushes asset, search for Biome Definition

4. Let's rename **BiomeName** to Bushes and change **BiomeColor** to a bright red color (**R**: 1.0, **G**: 0.0, **B**: 0.0, **A**: 1.0), as shown in the following screenshot.

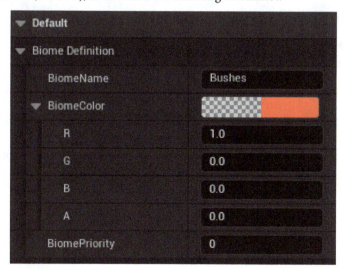

Figure 8.50 – Change BiomeColor to a bright red color – R:1.0 G:0.0 B:0.0 A:0.0

We just finished setting up the DA_BiomeDBushes data asset and now we can carry on and start setting up the PCG biome setup blueprint, but this time only to form the trees on the landscape! Let's get started:

1. Let's go back to the BushBiome folder, right-click on it, and from the menu, select **Blueprint Class**. Search for **BP_PCGBiomeSetup** and select it! Rename your blueprint to BP_BushBiome.

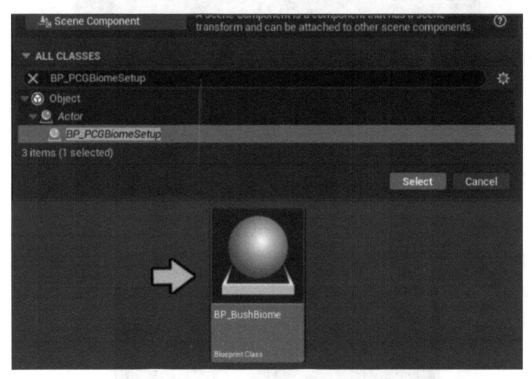

Figure 8.51 – Add a new actor blueprint, BP_PCGBiomeSetup, and rename it to BP_BushBiome

2. The first thing you will notice is that it's an actor blueprint. The only major difference is that it is programmed to only work with the DataAssets for PCG use.

Figure 8.52 – Open the BP_BushBiome actor blueprint

3.   On the **Details** panel on the left, choose **DA_BiomeDBushes** for **Definition**.

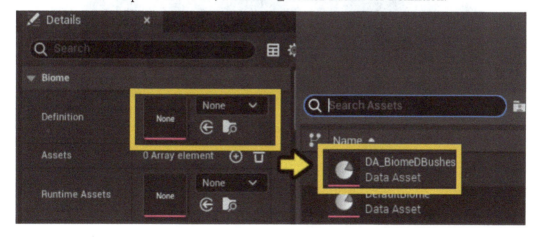

Figure 8.53 – On the Details panel, under Biome in the Definition
placeholder, add the DA_BiomeDBushes data asset

4. Just below **Definition**, in the **Assets** tab, add an array element. This will add an array element that will be used for data assets.

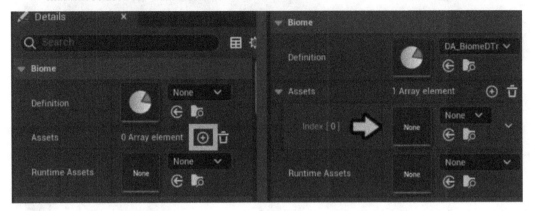

Figure 8.54 – Under the Assets panel, add an Index[0] array

5. For this index array, let's choose the **DA_BiomeABushes** asset.

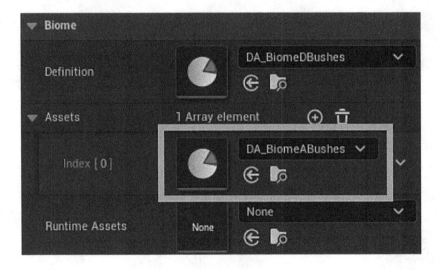

Figure 8.55 – In Index[0], assign DA_BiomeABushes

6. Let's compile the actor blueprint and return to the main viewport! Now let's drag and drop **BP_BushBiome** into the scene. Then, you should be able to see the trees start to appear in your scene – just on the green areas while you are in Debug mode.

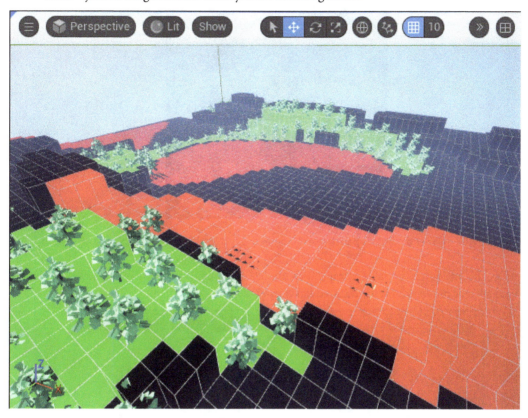

Figure 8.56 – A preview of the red color results on the landscape actor

The red blocks obstruct the view and hide the main bush models. It's the right time to turn off the debug mode. With the **BP_BiomeCore** blueprint selected, go to the **Details** panel and select the **BiomeCore** component. Then, turn off **Debug- Display Biome Cache** completely.

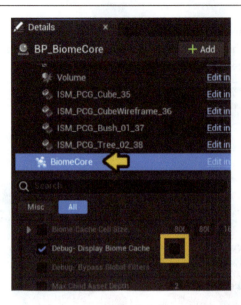

Figure 8.57 – Under the Details panel, select BiomeCore and disable Debug- Display Biome Cache

Here are the final results of the landscape, with meshes distributed across the color patches assigned to each specific model within the textured landscape.

Figure 8.58 – Final results of the PCG Biome on the Landscape actor

We've just reached the end of *Chapter 8*, and now it's time to put our knowledge to the test by experimenting with different static meshes for the foliage in the PCG biome.

Summary

In this chapter, you've picked up a handy trick that doesn't need a deep understanding of the PCG graph. You've learned how to use the latest PCG biome plugin for distributing foliage across the landscape. It's awesome to see how well you're progressing and how much you've achieved so far! You have already learned how to use the latest PCG Biome plugin, which is easier to work with compared to other PCG setups. Additionally, it allows you to customize the distribution of foliage based on the colors of your texture. This experience has also given you insight into using datasets for the first time within a PCG setup. Combining the PCG Biome and PCG Graph will further enhance the quality of your work.

Next up, we'll dive into animation. We'll show you how to bring animated characters or objects into the scene using the PCG framework.

9

Creating Dynamic Animated Crowds with PCG

Fantastic progress so far—you've just reached *Chapter 9*! In the last chapter, we explored the new Biome plugin, demonstrating how to create your own PCG biome by using data assets and evaluating them within the Biome plugin. This time, you'll learn how to spawn animated characters and transform them into a crowd of people! That's right: in this chapter, you'll gain insights into creating your own crowds with the help of the **PCG** plugin!

In this chapter, we will explore topics such as setting up actor blueprints for your characters, integrating PCG components into actor blueprints, and creating a PCG graph to support the character blueprint setup. We will use Mixamo animated models and demonstrate how to spawn characters within a closed spline loop, enabling each character to walk independently using simple AI components set up within each actor blueprint.

You'll gain hands-on experience with the advanced features of the PCG tool and learn how to utilize them while discovering additional solutions for creating believable crowd simulation experiences.

This chapter concentrates on the following key areas:

- Getting familiar with the Mixamo characters and animations
- Setting up the character actor blueprint
- Creating PCG actor blueprint
- Creating a PCG graph that will generate a crowd

By the end of this chapter, you'll have gained new techniques and valuable insights into setting up your own PCG crowd simulation.

Technical requirements

You will need the following hardware and software to complete this chapter:

- A good computer that can run a PCG, with the following specs: multi-core CPU (AMD Ryzen 7/9 or Intel i7/i9), GPU (NVIDIA RTX or AMD Radeon RX with 8+ GB VRAM), and at least 16 GB RAM.

- A basic understanding of setting up blueprints in Unreal Engine 5.

- The template project includes already prepared Mixamo models with the animations inside. Additionally, you can download Mixamo characters and set them up for your own purposes. But I advise you to follow the tutorial first with the examples prepared in this chapter!

For this project, we will use the template project, which is available in the GitHub repository, and you can download its template from GitHub at `https://github.com/PacktPublishing/Procedural-Content-Generation-with-Unreal-Engine-5`.

The code in action video for the chapter can be found at `https://packt.link/VSZOM`

Working with Mixamo characters

As we go along through this chapter, we'll be using the Mixamo characters that I have prepared for this exercise. You can also find more characters with animations at `https://www.mixamo.com/#`.

We will look at the Mixamo characters and examine each one, along with their animations.

Navigate to the `Mixamo` folder and, here, you will find folders for `Male` and `Female`, which have the animations and models for each character (see *Figure 9.1*). As you might have guessed, there are two characters: one male and one female. We will only use two characters for this exercise and we will create a Character Blueprint, which we will assign inside the PCG graph.

Figure 9.1 – List of Mixamo skeletal meshes with the animations

This concludes the section. In the next section, we will get into creating an actor blueprint for each character, and I will show you how to create the AI that will be suitable for each of these characters.

Developing a Mixamo character blueprint

In this section, we will get into setting up an actor character blueprint by simply creating a new character blueprint. We will create a method that selects a random skeletal mesh and then assigns a random animation to that skeletal mesh. Without further ado, let's get started:

1. In your `Blueprint` folder, right-click and select **Blueprint Class**. In the **Pick Parent Class** menu, select the **Character** blueprint. Rename your character blueprint `AI_Player`, as shown in *Figure 9.2*:

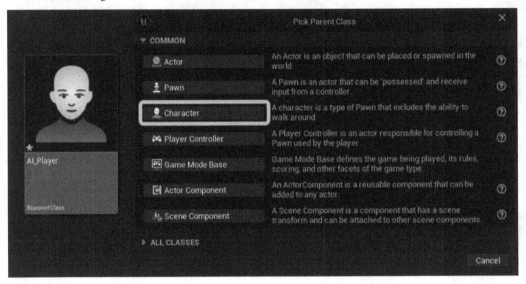

Figure 9.2 – Add the Character blueprint and rename it AI_Player

2. Open your `AI_Player` blueprint and go to the **Variables** section. Create two new variables and name them `SkeletalMeshes` and `AnimSequences`. For the `SkeletalMeshes` variable, set **Variable Type** to **Skeletal Mesh**, and for the `AnimSequence` variable, set **Variable Type** to **Anim Sequence**, as shown in *Figure 9.3*. Ensure you change their container type from **Single** to **Array**.

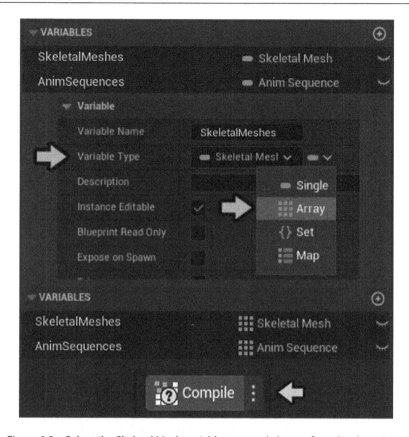

Figure 9.3 – Select the Skeletal Mesh variable type and change from Single to Array

3.   For each variable on the right-hand side, press the plus (+) button to add two array elements. We will use two of each of the skeletal meshes and the animation sequences.

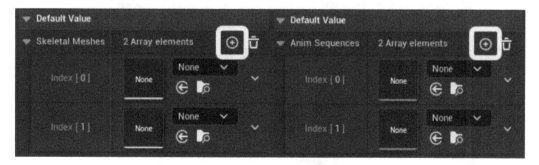

Figure 9.4 – Add two array elements for the Skeletal Mesh and Anim Sequence arrays

4. For the `SkeletalMesh` variable, let's add **BizMale** and **Remy**. For the `AnimSequence` variable, let's add the **Walking_M** and the **Texting_Idle** animations, as shown in *Figure 9.5*:

Figure 9.5 – Assign BizMale and Remy to the Skeletal Mesh indexes, and assign the
Walking_M and Texting_Idle animations to the Anim Sequence indexes

5. In your **Event Graph** space, right-click and look for the Custom Event node inside the *search* menu. Add that node and rename it `AIWalk`, as shown in *Figure 9.6*:

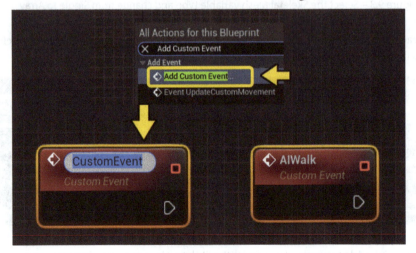

Figure 9.6 – Add CustomEvent and rename it to AIWalk

6. As you may have noticed, on the left-hand side, your player character has the **Character Movement** component, which is essential for your character to move in play mode. Drag and drop this component into the **Event Graph** area, placing it next to the **AIWalk** execution node.

Figure 9.7 – Drag and drop the Character Movement component to the Event Graph

7. From the **Character Movement** component, drag out the pin and search for the Set Max Walk Speed variable. Add it to the graph and place it close to your **AIWalk** execution node, as shown in *Figure 9.8*:

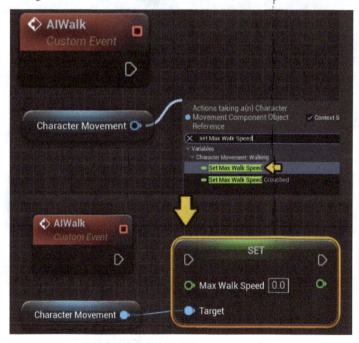

Figure 9.8 – Drag the pin out from the Character Movement component
and search for the Set Max Walk Speed variable

8. Let's change the **Max Walk Speed** in your setter to a value of 100.0. This should make the character walk at the desired speed; we will use this speed value for AI movement later. Connect it to the **AIWalk** execution pin:

Figure 9.9 – Set the Max Walk Speed to 100.0

9. The next step is to introduce the AI MoveTo function into this sequence. This will enable the character to detect the navigation volume (which we will set up later (see *Figure 9.61*), allowing it to roam around the environment.

Figure 9.10 – Add an AI MoveTo component to the Event Graph
and connect it to the Set Max Walk Speed variable

10. Let's add the **Get a reference to self** node and connect the **Get a reference to self** node to the **Pawn** input of the AI MoveTo function:

Figure 9.11 – Connect the Get a reference to self node to the Pawn input of the AI MoveTo function

11. Currently, your character will remain stationary without any set destination, meaning it won't do much. To address this, let's add two components that will determine the actor's location and a random location within its reachable radius in space. Right-click on the graph and search for the Get Actor Location and GetRandomReachablePointInRadius functions. Add both to the graph and connect them as shown in *Figure 9.12*:

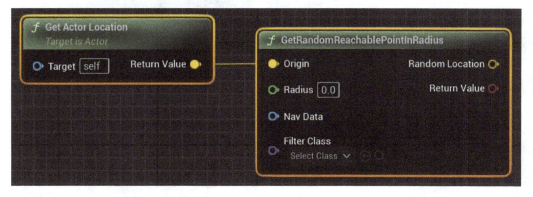

Figure 9.12 – Combine and link the Get Actor Location function with the
GetRandomReachablePointInRadius function using the Origin input

12. Set **Radius** for `GetRandomReachablePointInRadius` to `1000.0` and connect the function to the `AI MoveTo` **Destination** input:

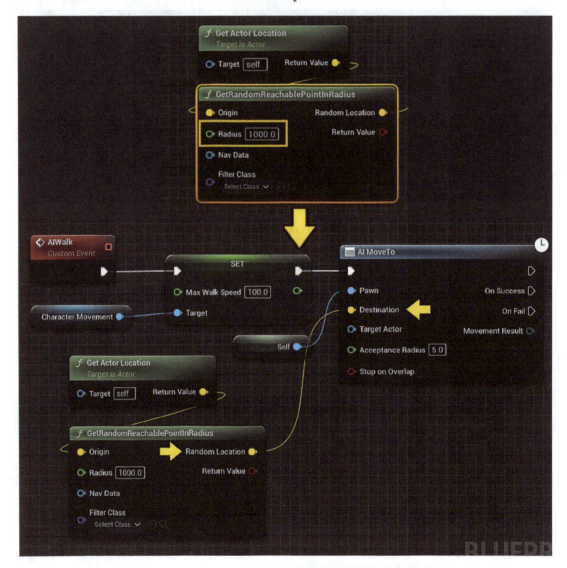

Figure 9.13 – Setting up the GetRandomReachablePointInRadius to the value of 1000.0 units

13. Inside the `GetRandomReachablePointInRadius` function, set its **Radius** to `1000.0` units. This ensures that the character will move within an area that is 1000.0 units away from its original location. Then, connect the **Random Location** output to the `AI MoveTo` function's **Destination** input.

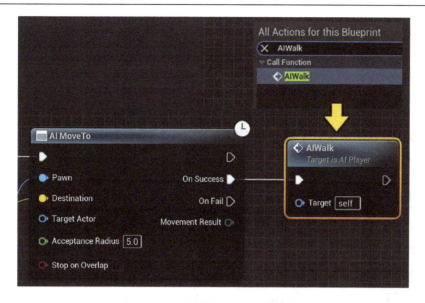

Figure 9.14 – Add AIWalk function and connect it to the OnSucess output of the AI MoveTo function

14. We are nearing completion. Now, let's loop this AI behavior by adding an **AIWalk** event at the end of this cycle. You can loop this behavior by using the **AIWalk** event as a separate function and calling it to repeat its process at the end of this chain. Right-click on the graph, search for the AIWalk function, then add it and connect the function node to the **On Success** output node of the AI MoveTo function. This will ensure that the AI movement process is managed correctly, preventing any potential looping issues later.

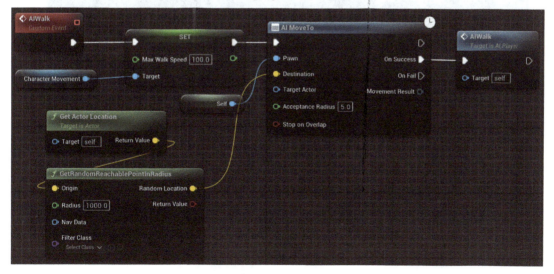

Figure 9.15 – Final look at the event graph with all the nodes connected

Figure 9.15 shows what your graph should look like with all the nodes connected!

The next step is to ensure it works in play mode. Before we set this up, we need to randomly spawn two different characters and assign animations to them randomly. In the next section, we will set up a random character spawning mechanism with various animations.

Interchanging skeletal meshes using the random integer function

In this section, we will configure the logic inside the Character Blueprint so that it will randomly swap skeletal mesh characters. This will introduce diversity to the spawned actors within the PCG graph once they are generated, allowing the crowd to represent different characters with randomly selected animations.

Now, let's dive into the step-by-step guide:

1. Select and drag out two arrays from the **VARIABLES** section: **SkeletalMesh** and **AnimSequence**. Make sure to set them as Get variables and add them to the Event Graph.

Figure 9.16 – Drag and drop the Skeletal Mesh and Anim Sequence arrays onto the Event Graph

2. The next step is to add two functions for each array variable to your event graph, and they are **Set Skeletal Mesh Asset** and **Play Animation**. As you may have noticed, the functions will already be assigned to the **Mesh** component, which helps to save you valuable time.

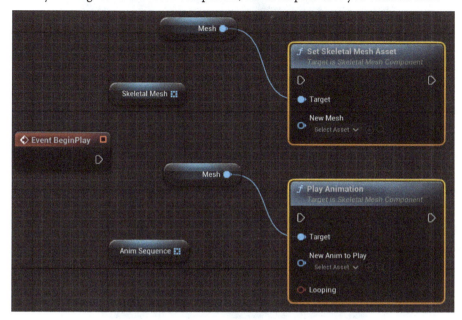

Figure 9.17 – Add the Set Skeletal Mesh Asset and Play Animation functions to the Event Graph

3. Let's get started with the **Skeletal Mesh** array variable. With the **Skeletal Mesh** variable selected, drag out the pin and search for the Get (a copy) node. Since this is an array, using the index 0 will retrieve the first element of the array. Hence, we will leave the value at **0**.

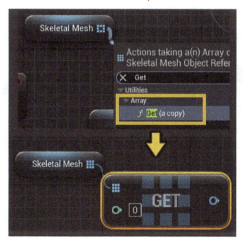

Figure 9.18 – Drag the pin out of the Skeletal Mesh array and search for Get (a copy) node

4.   From the blue **GET** node output pin, connect to the **New Mesh** input node in the **Set Skeletal Mesh Asset** function. Connect **Set Skeletal Mesh Asset** directly to **Event BeginPlay**.

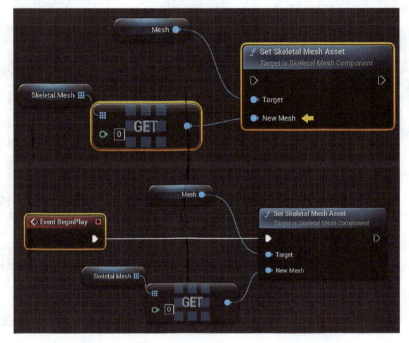

Figure 9.19 – Connect the Skeletal Mesh Array to the New Mesh input of the SetSkeletalMeshAsset function. Then, link the Event BeginPlay node to the SetSkeletalMeshAsset function.

5.   Before we proceed to the **Play Animation** function, let's set up a simple condition to randomly select an animation and assign it to the spawned skeletal mesh in the scene. Right-click on the graph and search for the Branch Flow Control node.

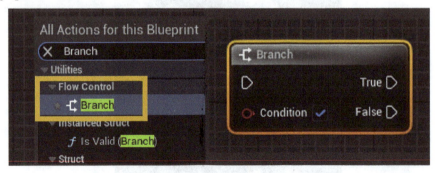

Figure 9.20 - Adding Branch Flow Control node

6. Let's add two more nodes: **Random Integer in Range** and the **Equal** sign operator. The **Random Integer in Range** node will be used to select a number between the minimum and maximum values of the array. The **Equal** sign operator will then check if the randomly chosen number matches the number specified within the array.

Figure 9.21 – Add Random Integer in Range function and the Equal numerator to the Event Graph.

7. With the nodes set for this exercise, let's connect them all together to form a logical condition for the skeletal mesh and the animation spawner. Connect the **Return Value** of the **Random Integer in Range** node to the first slot position of the **Equal** sign operator.

Figure 9.22 – Connect the Return Value of the Random Integer in
Range to the first input of the Equal numerator

8. Ensure that the **Min** value is set to 0 and the **Max** value is set to 1 on the **Random Integer in Range** node. Then, set the second slot of the **Equal** operator to 0. This condition will determine that if the first skeletal mesh is spawned, one of the two animations will be assigned to the newly spawned skeletal mesh. Connect the **Condition** to the **Branch**'s Condition value input.

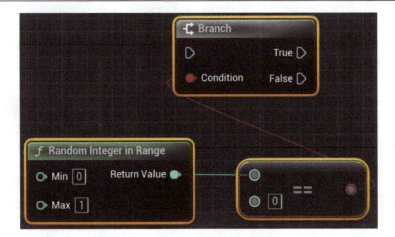

Figure 9.23 – Connect the Equal Numerator to the Branch condition

9. We can now put that **Condition** at the top of the graph and place it next to the **Set Skeletal Mesh Asset** function to continue the flow of our **Event BeginPlay** execution node. Connect it to the existing nodes.

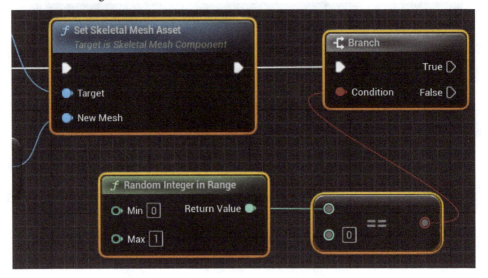

Figure 9.24 – Connect the Branch to the Set Skeletal Mesh Asset function

10. To complete this part, connect the **Random Integer in Range** function node to the **GET** node, which is already connected to the **Skeletal Mesh** array variable at the beginning of our graph. This will ensure that we are first getting the mesh and then getting another random integer to get one of the two animations.

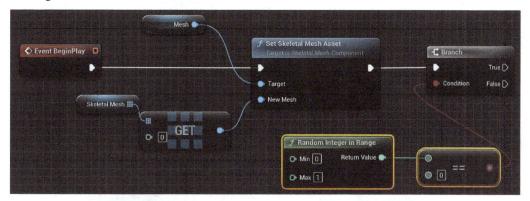

Figure 9.25 – Final look at the node connections to Event BeginPlay

11. Now, let's connect the **Random Integer in Range** node to the **GET** node of the **Skeletal Mesh** array variable, linking it to the integer input.

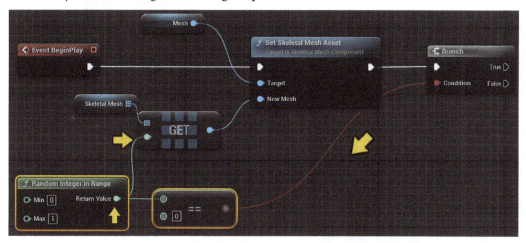

Figure 9.26 – Connect the Return Value of the Random Integer in Range function to the GET node

12. Next, we will set up a **Play Animation** to play one of the animation sequences. With the **Anim Sequence** array variable selected, drag the pin out and add a **Get (a copy)** node.

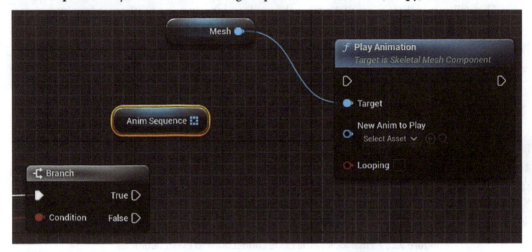

Figure 9.27 – Select the Anim Sequence node

13. Connect the **GET** node to the **New Anim to Play** input of the **Play Animation** function.

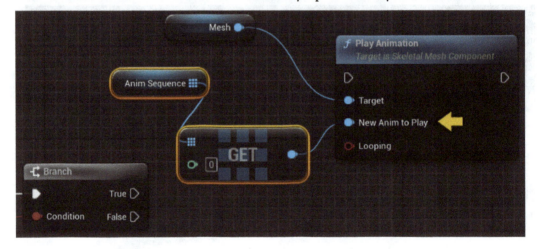

Figure 9.28 – Retrieve the node from the Anim Sequence variable and link
it to the New Anim to Play input of the Play Animation function

14. Now, connect the **Play Animation** function to the **Branch** node's **True** statement output. Set the **GET** node to read index 1 value. This means that if the random integer is set to 1, it will play an animation that is stored in the array at index 1 in the **Anim Sequence** array variable. Make sure to enable the **Looping** mechanism inside the **Play Animation** function.

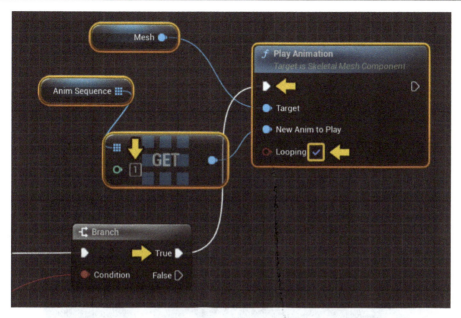

Figure 9.29 – For the True statement, set the array value to 1 on the GET node,
and activate the Looping boolean within the Play Animation function

15. Let's copy the entire **Play Animation** function graph and paste it below the **Branch** node. This time, connect the duplicated nodes to the **False** statement.

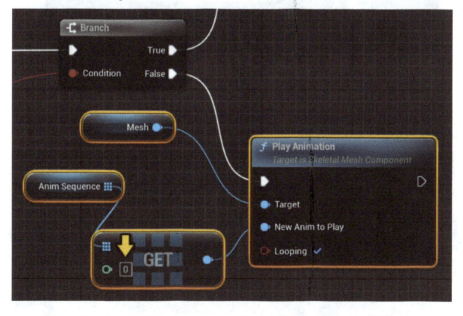

Figure 9.30 – For the False statement, set the array value to 0 on the GET node,
and activate the Looping boolean within the Play Animation function

This means that if the number of the array is set to 0, it will not play the animation from index 1 but will instead play the animation from index 0 of the **Anim Sequence** array variable. Ensure that the **GET** node is set to an array value of 0.

16. The last part is to ensure that when the walking animation is selected, the AI character continues to roam and walk around the terrain. As you can see in the **Anim Sequence** variable, the walk cycle animation is set in the first index of the array variable (see *Figure 9.31*):

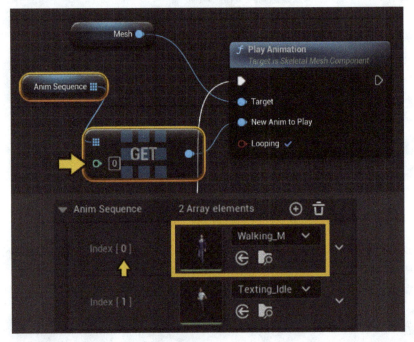

Figure 9.31 – Indexes corresponding to each Anim Sequence animation asset

17. Let's add the **AIWalk** function at the end of the play animation function, where the **Anim Sequence** variable with a **GET 0** index is assigned to the **False** statement.

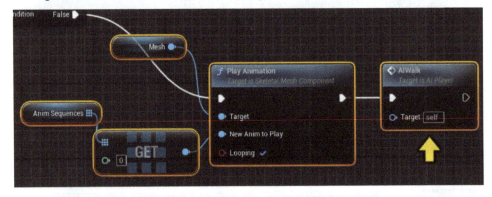

Figure 9.32 – Add the AIWalk function and connect to the Play Animation function

You have achieved your first character blueprint that will work later within a PCG graph and it will respond with its own AI movement.

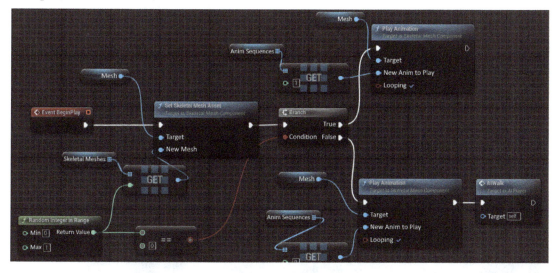

Figure 9.33 – Final look at the Even Graph nodes structure

In the following section, we will focus on incorporating the remaining elements for the character setup.

Setting up the character

The last part of the puzzle is to make sure we have a reference of the skeletal mesh attached to the character. It's crucial to add the character as a skeletal mesh so we can easily see its spawning position later. In this case, we have properly placed the character mesh so that it is pointing in the right direction. Let's go ahead and add the Mixamo character to the character mesh.

Select your **Mesh** component from the **Components** window in the left top corner. Then navigate to the **Mesh** tab on the right-hand side, search for the `BizMale` skeletal mesh, and select it.

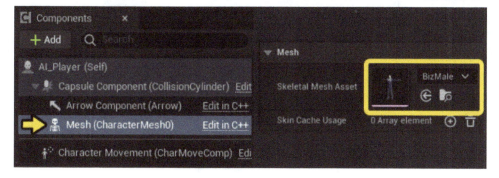

Figure 9.34 – Select the Mesh (CharacterMesh0) component and assign
the BizMale skeletal mesh to the Skeletal Mesh Asset

In the Viewport window, you may have noticed that the character is a little bit offset, and it needs some attention. We have to make sure that the characters will be pointing in the direction of the *blue* arrow reference that is placed inside the capsule.

On the left-hand side, change its translations on the Z axis (colored *blue*) to the value of minus 80.0 units and then turn the character around by the minus 90.0 degrees around the Z axis (Yaw).

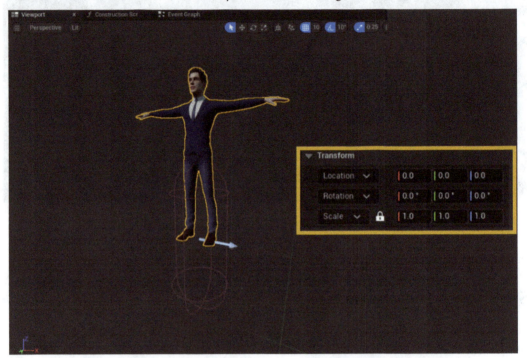

Figure 9.35 – Inspecting the Transform settings for the Skeletal Mesh

Take a look at the following table for better readability and copy the following values:

Location	0.0	0.0	-80.0
Rotation	0.0	0.0	-90.0
Scale	1.0	1.0	1.0

Table 9.1 – Transform settings

We have now updated the **Transform** values for our character mesh using the values from the preceding table.

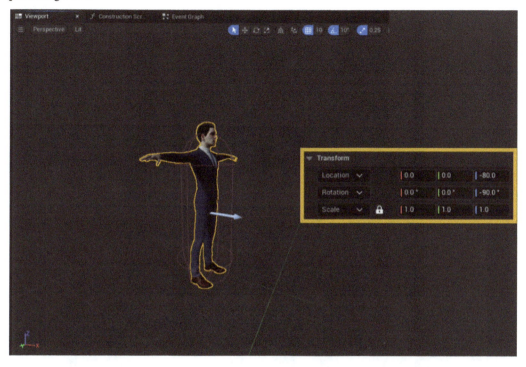

Figure 9.36 – Changing the direction of the skeletal mesh to 90 degrees on the Z axis

During gameplay mode, your character will rotate very quickly about its axis without a smooth transition, which means it doesn't calculate the rotation properly, It skips the smooth transition as it quickly twists around its axis. To fix this issue we need to tick a few boxes inside the character blueprint:

1. Select the top main **AI_Player** component inside the **Components** window. On the right-hand side, on the **Details** panel in the top search bar, type in Use Controller Desired Rotation.

2. Tick the box to enable it. This will allow the character to rotate in any direction within the minimum and maximum angle.

Figure 9.37 – Select the AI_Player and go to the Details panel.
Search and enable Use Controller Desired Rotation

3. Let's check another setting: in the search bar, type in Use Controller Rotation Yaw. Make sure to tick the box to disable it completely. It will turn off the snapping rotation around the character's axis.

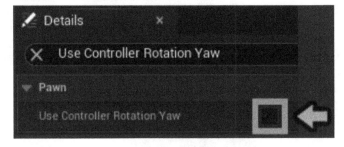

Figure 9.38 – Search and disable Use Controller Rotation Yaw

Now, you can compile the character blueprint and save it. You have just completed an important part of the practical tutorial: setting up the character mesh with the functionality of walking by itself using a simple AI functionality that's built into Unreal Engine 5!

In the next part of this chapter, we will prepare the actor blueprint with a spline controller that will hold the PCG graph inside it with a few variable controls for the number of people spawned in the crowd.

Setting up the PCG crowd blueprint

In this section, we will set up an actor blueprint, which will spread the number of **AI_Player** characters within the radius of the spline controller. It will be a very easy setup and it won't take too long!

Now, let's dive into the step-by-step implementation:

1. Inside the `Blueprint` folder, create an **Actor** blueprint and call it `BP_PCGCrowd`.

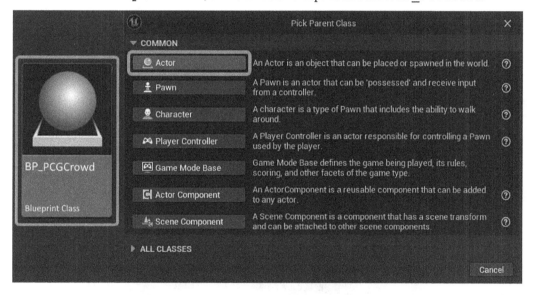

Figure 9.39 – Create a new Actor blueprint and rename it to BP_PCGCrowd

2. Let's open your new actor blueprint and, on the **VARIABLES** tab, add a float variable and call it `SpawningDistance`. This will control the distance between the distribution of newly spawned **AI_Player** actors. Be sure to always makethe variables public.

Figure 9.40 – Add a new float variable and rename it to SpawningDistance

3. With the **SpawningDistance** variable selected, in the **Details** panel on the right-hand side, add the **Slider Range** and the **Value Range** starting from 100.0 and closing at 500.0. Make sure to set the **Spawning Distance** variable's value to 100.0 in order to prevent any unpleasant crashes while working with the PCG graph later. Hit **Compile** and save it.

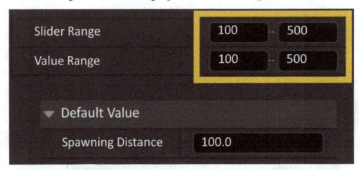

Figure 9.41 – Set the slider and value range to 100 and 500 respectively

4. In the **Components** panel, add a **Spline**.

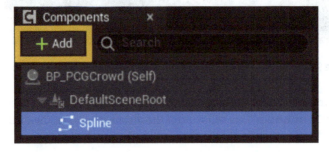

Figure 9.42 – Add a Spline component

5.   With the **Spline** controller selected, inside the **Viewport** menu, right-click on the gizmo selection and, in the menu, choose **Spline Generation Panel**.

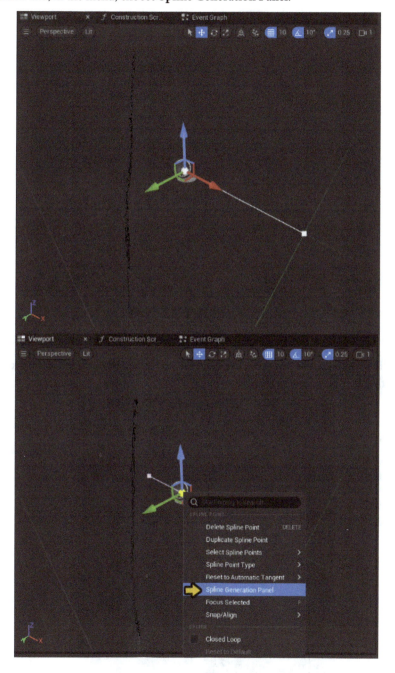

Figure 9.43 – Right-click on the spline point and search for the Spline Generation Panel

6.   Inside the **Spline Generation** panel, select **Circle**. With the current settings, you may now close the window.

Figure 9.44 – In the Spline Generation panel, choose Circle

7.   Your spline will have the first array selected. Hit the *delete* key button on your keyboard to remove that first index spline point.

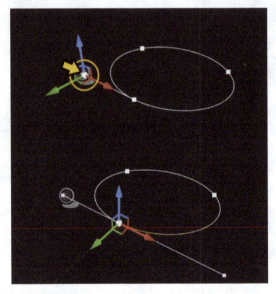

Figure 9.45 – Remove the first spline point

8.   Make sure to tick the **Closed Loop** setting on the **Details** panel.

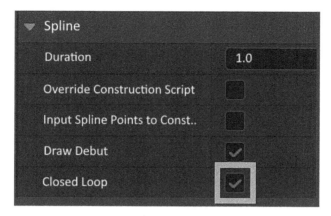

Figure 9.46 – In the Spline panel, enable Closed Loop

9.   Lastly, let's add the **PCG** component to the **Components** panel. For now, we don't have the PCG graph ready, but we will come back to the actor blueprint toward the end of this chapter.

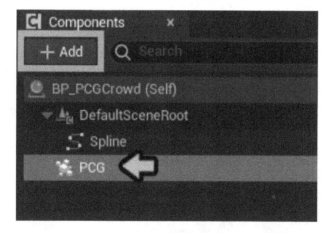

Figure 9.47 – Add PCG component

10. Compile and save your actor blueprint. Let's close it and drag and drop it into the scene for now.

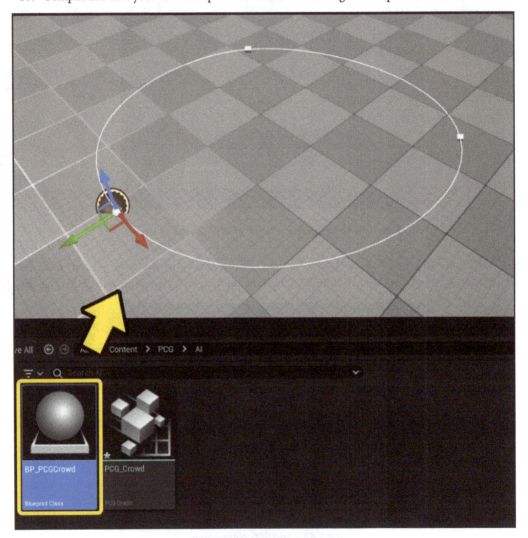

Figure 9.48 – Drag and drop the BP_PCGCrowd actor blueprint into the scene

You have just finished setting up the actor blueprint for the upcoming PCG crowd. In the next section, we will focus on creating the PCG graph and start setting up the graph structure that will allow us to spawn all the character mesh actors inside the spline controllers.

Setting up the PCG crowd graph

We just reached the final part of this chapter, and we will create a PCG graph that is capable of spawning actor characters across the level. I will show you how to set up a logic system for spawning the actors from the PCG graph:

1. Inside your PCG folder, right-click and create a new PCG graph, and let's call it PCG_Crowd.

2. Open your graph, and let's start adding the nodes to the PCG graph! The nodes we are going to add are: **Get Actor Data**, **Spline Sampler**, **Transform Points**, and **Spawn Actor**. The last node generates and spawns all objects associated with the actor class.

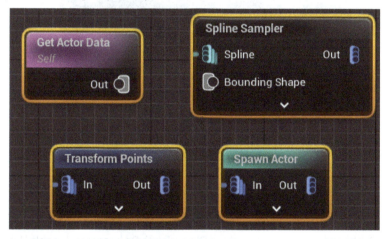

Figure 9.49 – Add Get Actor Data, Spline Sampler, Transform Points, and the Spawn Actor nodes to the PCG graph

3. Firstly, connect the **Get Actor Data** node to **Spline Sampler** and then **Spline Sampler** to the **Transform Points** node.

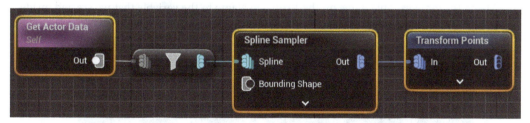

Figure 9.50 – Connect the nodes

4. Select the **Spline Sampler** node and, on the right-hand side, on the **Settings** panel, change **Dimension** mode to **On Interior** mode. On the same **Settings** panel, scroll down and make sure to check the **Unbounded** box.

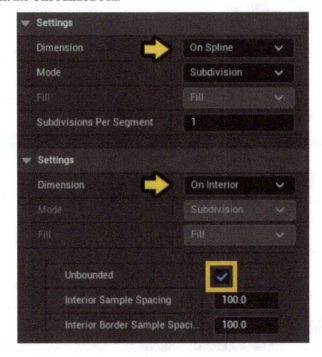

Figure 9.51 – On the Spline Sampler, change the Dimension to On Interior and enable Unbounded

5. Let's add one more node, **Get Actor Property**, which will be our variable that will be connected to the variable created in the **BP_PCGCrowd** actor blueprint. It will be used to control the spawning distribution within the spline interior.

6. Match the same variable name on the **Property Name** to the variable that we created inside the **BP_PCGCrowd** blueprint, and that is **SpawningDistance**.

Figure 9.52 – Add Get Actor Property and change the Property Name to SpawningDistance

7. Connect the **Get Actor Property** with the **Spline Sampler** node's **Interior Sample Spacing** input, which you can find at the bottom of the list on the **Spline Sampler** node, which you can find when you click on the down arrow.

Figure 9.53 – Connect the Get Actor Property Spawning Distance node to Interior Sample Spacing

8. Select **Transform Points** and add a few transformation values for the rotation and the scale parameters. It will help to slightly differentiate the character's heights and sizes.

Figure 9.54 – In the Transform Points node, adjust the values

For additional details, copy the transform values from the following table:

Rotation Min	0.0	0.0	-90.0
Rotation Max	0.0	0.0	90.0
Scale Min	0.8	0.8	0.8
Scale Max	1.0	1.0	1.0

Table 9.2 – Transform Points values

9.   Now, it's time to add the actors to the scene through the PCG graph. Select the **Spawn Actor** node. On the right-hand side, in the **Settings** panel, change **Option** to **Merge PCGOnly**. This will allow spawned actors to work within the PCG.

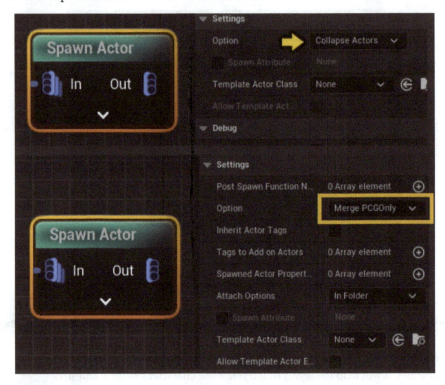

Figure 9.55 – In the settings of Spawn Actor, change Option from Collapse Actors to Merge PCGOnly

10.   The last part is to assign the right actor under the **Template Actor Class**, and to do this, we will use our created **AI_Player** character actor.

Figure 9.56 – Under Template Actor Class, search for and choose AI_Player character blueprint

11. Let's connect the **Spawn Actor** node to the **Transform Points** node to finish the PCG graph node structure.

Figure 9.57 – Finally, connect the Spawn Actor to the Transform Points node

12. Now it's the right time to test our PCG graph. To do that, we have to go back and open our **BP_PCGCrowd** actor blueprint. Select the **PCG** component in the **Components** window and assign the **PCG_Crowd** graph on the **Instance** panel on the right-hand side.

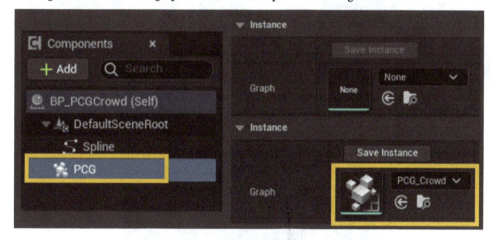

Figure 9.58 – Select the PCG component and, on the Instance tab, add the PCG_Crowd graph

13. Hit the **Compile** button and save your progress. Now, you can give it a go and test the actor blueprint on the scene. The character will perform some animations that we have assigned, but they won't walk around because we have to add more missing assets to the scene.

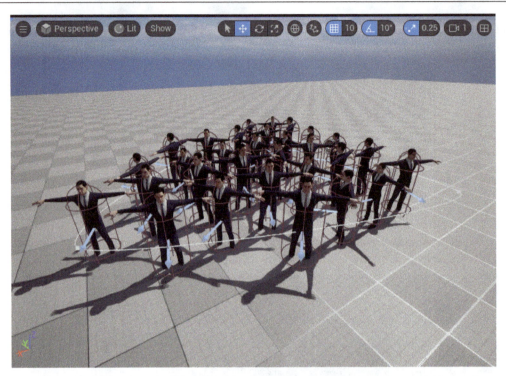

Figure 9.59 – Preview of the working BP_PCGCrowd actor blueprint in the scene

14. On the top left of the viewport window, click on the plus (+)button and search for Nav Mesh Bounds Volume. Add it to your scene and resize it to cover the entire floor.

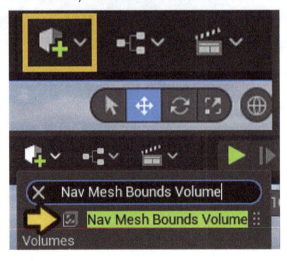

Figure 9.60 – Searching for Nav Mesh Bounds Volume

15. After resizing your **Nav Mesh Bounds Volume**, if you want to see debug mode, press the *P* key button to visualize the coverage of the entire floor. You should see everything covered in a *green* transparent color. This indicates that the navigation covers the sections within the volume space where characters will be roaming, thanks to the AI functionality inside **BP_PCGCrowd** ...blueprint!

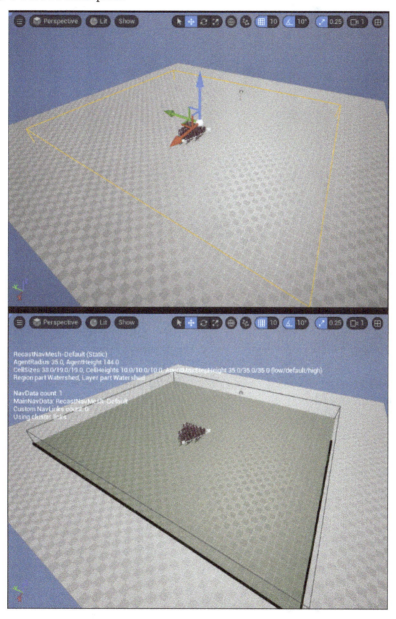

Figure 9.61 – Configuring the Nav Mesh Bounds volume to work with the BP_PCGCrowd actor blueprint

Now, let's dive in and check the results in the main scene. You will see characters going about their business on the level.

Figure 9.62 – Final results of the working BP_PCGCrowd actor blueprint in the scene

You have reached the end of this chapter, and you can give it a go and see your blueprint in action by pressing the play button. Make sure to press the *P* button again to turn off debug mode.

Summary

We have completed *Chapter 9*, and you have gained valuable insights into creating custom actor blueprints and integrating them within the PCG graph. This chapter has shown you how to effectively combine actor blueprints with the PCG plugin.

You now have a solid understanding of the skills needed to build a functional PCG for simple crowd simulations, enabling you to test and create different types of crowds repeatedly. Feel free to create your own character mesh characters using the template you prepared in this chapter!

In the next chapter, I will guide you through optimization and debugging techniques within the PCG graph. You will learn how to maximize performance using the PCG graph, reduce the number of draw calls, and manage heavy geometry and assets in Unreal Engine 5.

Part 3:
Mastering Optimization and Elevating Your PCG Environments

In this section, as we near the end of the book, we'll dive into optimizing and fine-tuning the PCG environment tools you've created. You'll discover solutions to boost the performance of your levels and complex environments, ensuring they run smoothly. We'll also explore creative techniques for adding extra nodes that enhance the visual appeal of your PCG environments, with these enhancements included as practical shortcuts.

This part has the following chapters:

- *Chapter 10, Exploring Optimization, Debugging, and Performance Tools*
- *Chapter 11, Cheat Sheets, Extra Tips, and Shortcuts*

10

Exploring Optimization, Debugging, and Performance Tools

It's been a great and intensive journey throughout this entire book. We have now reached *Chapter 10*, which is primarily dedicated to performance optimization. This chapter will guide you on how to further enhance the fluidity of the PCG tool within your project.

We will take some time to discuss how to get your project off the ground using PCG and how to achieve an amazing PCG environment without encountering performance issues. We'll walk through various examples and methods designed to improve project performance and provide essential solutions for working with assets that will be procedurally generated using your PCG tools.

In this chapter, we will use simple examples from the `Plugin` folder and discuss how to further optimize Megascan Quixel assets. This chapter will provide valuable insights into making assets work efficiently with any PCG tools.

In this chapter, we take a look at the following topics:

- Exploring different optimization techniques
- Exploring debugging tools
- Performance tools

Technical requirements

You will need the following hardware and software :

- A good computer system that can run a PCG; those are as follows:
- Ideally, it should have a multi-core CPU (AMD Ryzen 7/9, Intel i7/i9), GPU (NVIDIA RTX, AMD Radeon RX with 8+ GB VRAM), and at least 16 GB RAM.

Along with these, this chapter expects the reader to have a basic understanding of setting up blueprints in Unreal Engine 5.

For this chapter, we will use a template project from the PCG `Plugin` folder (**Engine** | **Plugins** | **Procedural Content Generation Framework (PCG) Content**| **SampleContent** | **SimpleForest** | **Meshes**). Additionally, we will use the **Trees: European Beech** asset from the Megascans European Trees package.

The code files for the chapter are placed at `https://github.com/PacktPublishing/Procedural-Content-Generation-with-Unreal-Engine-5`

Exploring different optimization techniques

In this section, we will explore a different optimization technique that can be implemented within the PCG framework. We will start by focusing on geometry and its properties, using this understanding to introduce various options for enhancing the project's framerate performance. Mesh geometry plays a crucial role in the PCG plugin in Unreal Engine 5.

Optimization, complexity, and performance

Using simpler meshes can improve performance as they require fewer calculations. This is ideal for large-scale procedural generation where performance is a concern. With a simple mesh, you can achieve up to 120 FPS, ensuring stable performance throughout the entire gameplay. The following is an example where the PCG uses foliage meshes with the lowest polycount polygons.

Figure 10.1 – Using a simple mesh for the PCG setup

Figure 10.2 shows the basic meshes available in the `Plugins` folder for the PCG framework. These models are ideal for enhancing your gameplay performance.

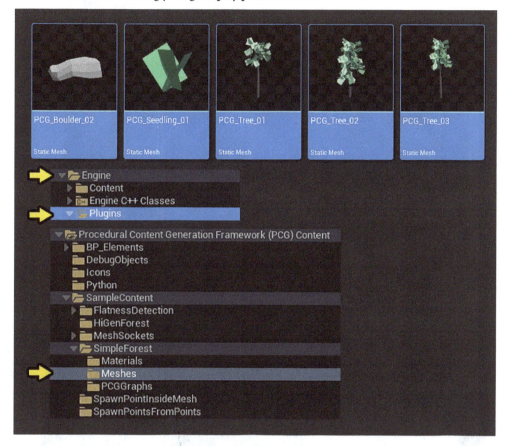

Figure 10.2 – Simple mesh models

More intricate meshes can add visual richness but may increase the computational cost, as you can see in the *Figure 10.3* example. When we replaced all the simple geometry models with Megascan tree assets, the performance dropped to around 70 FPS. This significant decrease can impact the overall gameplay experience, especially when using photorealistic, high-polycount assets. Balancing complexity and performance is key.

The following illustration represents the same PCG volume but it is replaced with the Megascan assets:

Figure 10.3 – PCG Volume working with the Megascan forest

Here are the Quixel Megascan assets that are used for the PCG volume example presented in *Figure 10.3*:

Figure 10.4 – Unreal Marketplace Tree Megascan Assets

In the next step, we will talk about collision and how collisions can be simplified to boost performance.

Collision

Accurate collision meshes are essential for gameplay mechanics, ensuring characters can interact with the environment properly. High-detail collision meshes can impact performance, so using simplified collision meshes is often beneficial. We can adjust the collision settings to enhance the performance of the meshes generated in the scene.

While inspecting the model in the static mesh viewport, switch the view mode to **Visibility Collision** to make the collisions visible.

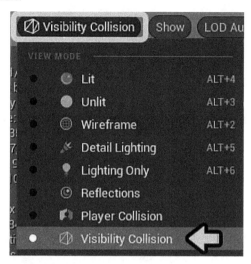

Figure 10.5 – Changing the Lit mode to the Visibility Collision mode

Figure 10.6 shows the difference between the complex and simple collisions.

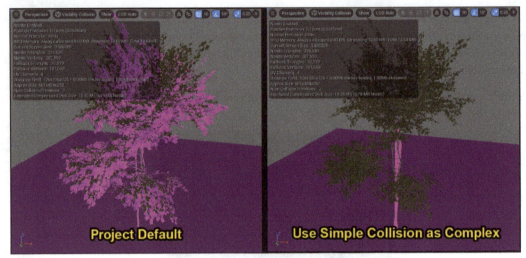

Figure 10.6 – The difference between complex and simple collisions

It is a good idea to change collision settings and make them less costly by setting **Collision Complexity** from **Project Default** to **Use Simple Collision As Complex.**

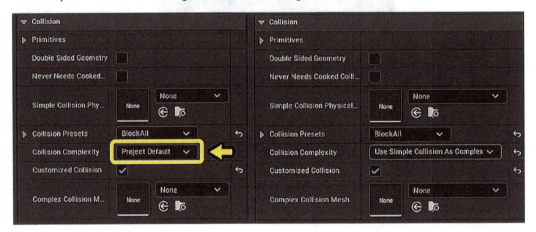

Figure 10.7 – Changing the collision complexity settings to a simpler collision

In the following section, we will explore the concept of modular pieces and how to utilize them in your project to maximize efficiency when using the PCG framework.

Modularity and reusability

Designing modular mesh pieces allows for more flexible and varied procedural generation. These pieces can be combined in multiple ways to create diverse environments. Reusable meshes save time and resources by fitting into various procedural scenarios with generic yet versatile geometry. For example, the Megascan asset illustrates a typical modular piece: the default version on the left in *Figure 10.8* has the pivot at the left corner, while the modified version on the right has the pivot offset to the middle. Both versions work fine but remember to calculate the spawning position of the modular mesh piece within the PCG graph.

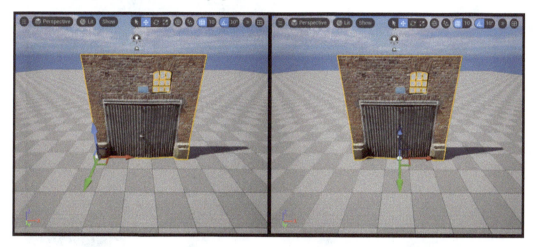

Figure 10.8 – Changing the pivot position for the modular static mesh

There is a certain trick that you can do in Unreal Engine 5 if you want to offset the pivot of the given mesh. I will illustrate this for you in the following example:

1. With your model selected, go to the top of the viewport, and under **Selection Mode**, choose **Modeling**.

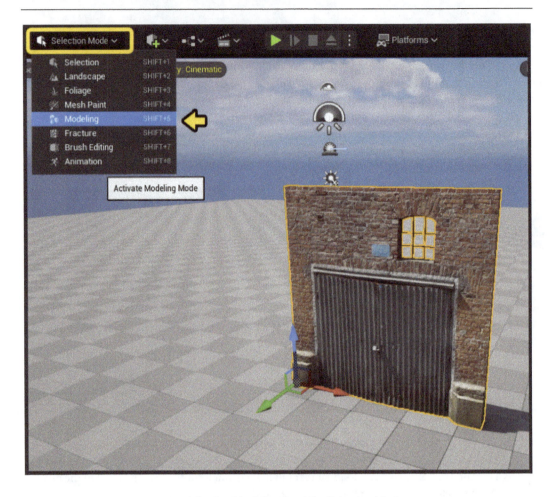

Figure 10.9 – Selecting Modeling from the Selection Mode menu

2. On the left panel, select the **XForm** tab, and inside that, select **Edit Pivot**.

Figure 10.10 – Selecting the Edit Pivot functionality under the XForm panel

3. This will enable pivot mode, which you can adjust to suit your needs. In this case, we are moving the pivot toward the middle by moving it 200 units to the right t! You can also select the red line of the pivot gizmo and move it manually, which is shown in the following figure.

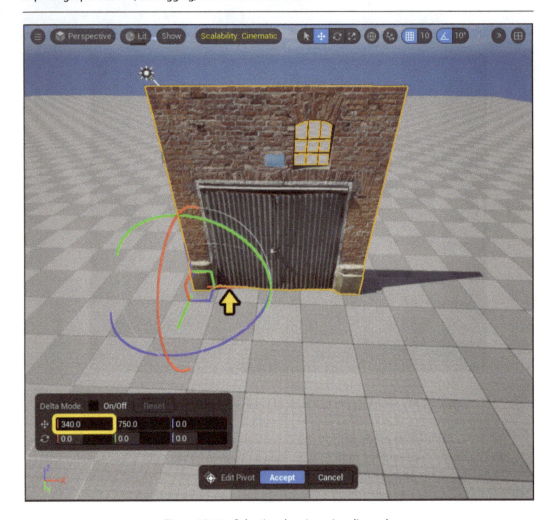

Figure 10.11 – Selecting the gizmo in edit mode

4.   Once you have moved the gizmo to the middle position, press the **Accept** button at the bottom of the screen. This will automatically assign the pivot to a middle position in your model.

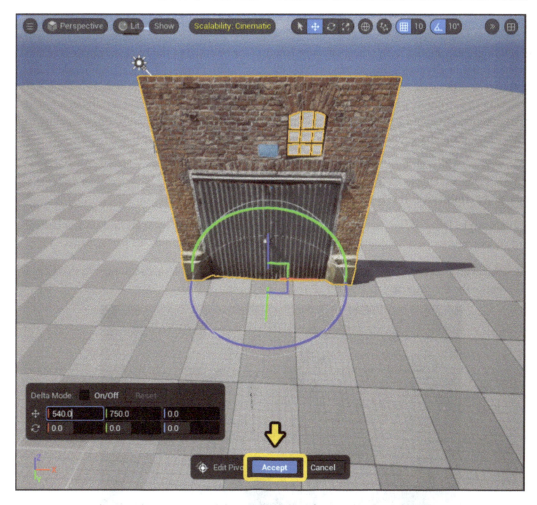

Figure 10.12 – Adjusting the pivot toward the center and confirming the changes with the Accept button

With these methods, you can save time, avoid using other DCC programs to adjust the pivot, and re-import the model into Unreal.

In the next section, we will explore the importance and use of the Nanite virtualized geometry system and how it can enhance the workflow of rendering spawned static meshes.

Nanite

As you might expect, this is the most effective solution for optimizing the large number of meshes spawned procedurally across the landscape. Typically, such optimization techniques require additional effort to reduce draw calls, which can slow down performance. However, Nanite enables the efficient rendering of extremely detailed models with millions of polygons. In procedurally generated environments, which often feature numerous unique assets and intricate details, this capability to manage high geometry counts without compromising performance is essential.

When working with high-quality assets, it's crucial to inspect the model to ensure it performs effectively across larger areas. Due to their mesh complexity, these assets can increase the number of draw calls, impacting performance. This is where Nanite proves to be a valuable solution.

As an example, let's take an asset from the Megascan library, which you can find on the Unreal Marketplace, and inspect its quality to understand how Nanite can optimize its performance:

1. Select an asset and choose a high-quality tree model from the Megascan library. In this case, I chose the `SM_EuropeanBeech_Field_01` mesh.

Figure 10.13 – Locating the SM_EuropeanBeech_Field_01 static mesh

2. Double-click on your tree static mesh. Inside the model, on your right-hand side panel under **Nanite Settings**, make sure to tick **Enable Nanite Support**. This will help you to enable Nanite for this particular model.

Figure 10.14 – Enabling Nanite Support under Nanite Settings

3. You can also enable Nanite for all models simultaneously. To do this, select all the models within the same folder, right-click on any of the selected models, and click on **Enable Nanite**. This will enable Nanite for all the selected meshes.

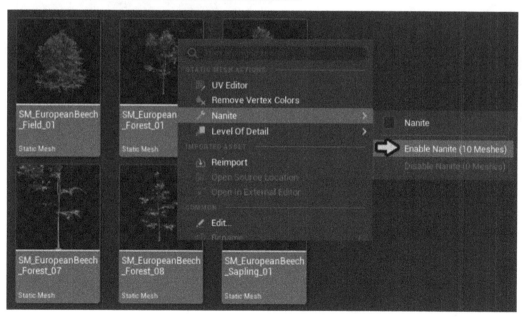

Figure 10.15 – Enabling Nanite for all the selected objects

In the following section, we will explore additional possibilities for optimizing the instance static meshes.

Creating a PCG Stamp

The PCG system offers another useful feature that helps maintain performance flow within the scene. This function eliminates the need for further use of the PCG volume inside the level by creating a **PCG Stamp**. The PCG Stamp packages all the PCG data and removes all the links from the volume that continually generates the meshes, optimizing the process. Basically, it lets you avoid regenerating every time you move the PCG volume around the level. It's up to you whether you want to stick with the current results or generate new changes within the volume, without needing to create a PCG Stamp. Let me show you an example here by using the project example from *Chapter 6*:

1. I have selected an actor blueprint called `BP_PCGLandscape`, which consists of the PCG component under the **Details** panel.

Figure 10.16 – Selecting a PCG component inside BP_PCGLandscape

2. With the **PCG** component selected, under the **PCG** tab panel, click on **Clear PCG Link**.

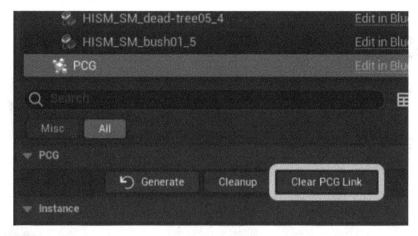

Figure 10.17 – Pressing the Clear PCG Link button

3. This method will create a PCG Stamp actor, which will be visible on the outliner. With this PCG Stamp, I can easily remove the original BP_PCGLandscape actor from the outliner while only keeping the PCG Stamp on my scene.

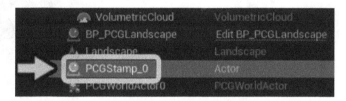

Figure 10.18 – Creating a PCGStamp actor

Now, with the PCGStamp actor selected, you can observe the performance difference it makes in your scene.

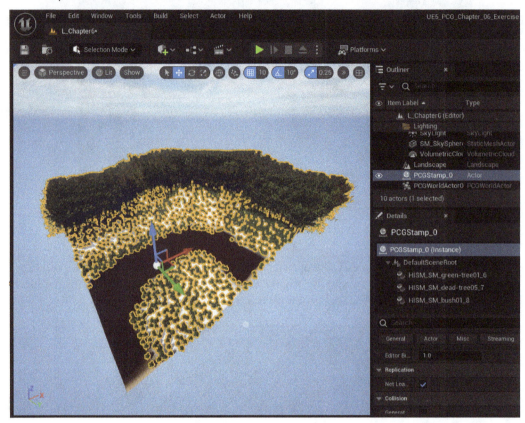

Figure 10.19 – Checking the results of a newly created PCGStamp

In this section, we will delve into the hidden optimization settings within the static mesh spawner, which further enhance the efficiency of spawning meshes.

Static mesh spawner

The **World Position Offset** (**WPO**) in the static mesh spawner node of a PCG graph is a crucial setting for optimizing performance in Unreal Engine. It allows us to disable vertex animations or offsets at a specified distance from the camera, reducing the computational load and improving performance without sacrificing visual quality where it matters most. Here is a practical example:

1. Inside your PCG graph, ensure that you have the **Static Mesh Spawner** node selected. Under the **Mesh Selector** panel, under **Mesh Entries** | **Index[]** | **Descriptor**, scroll down until you find

the **Evaluate World Position Offset** parameter. At the moment, it is enabled by default, but you can uncheck the box to disable it. When disabled, any material relying on World Position will be deactivated, affecting elements like moving tree branches, grass, wind, and leaves s. This will help improve the performance by even an additional whopping 10-20 frames per second.

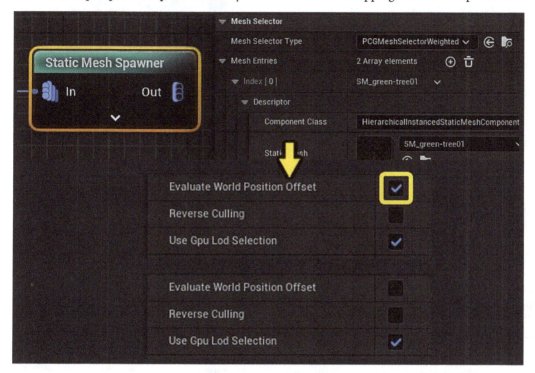

Figure 10.20 – Disabling the Evaluate World Position Offset setting

2. Go back to your main viewport and change from **Lit** mode to **Virtual Shadow Map | Cached Page** mode (*Figure 10.21*). Then you should be able to see a difference, especially when you turn on the frame rate on. Green represents it being turned off and blue represents it being turned on!

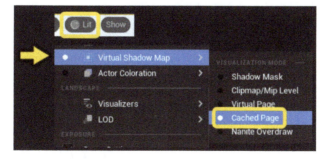

Figure 10.21 – Setting up the Cached Page vizualisation mode

Two configurations enable the observation of differences between visible and non-visible WPOs, which are displayed using green and blue coloration.

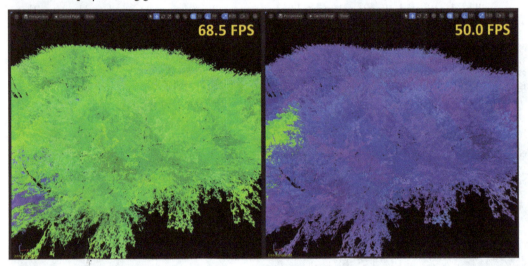

Figure 10.22 – Visible WPO distance ranges

In the next section, we will delve into configuring the WPO distance. I will guide you through how to work with this setting within the **Static Mesh Spawner** settings panel.

Enabling the WPO distance

Another approach to using World Position is to enable it but use it to reduce the effects of the WPO depending on how far you are from the foliage. In this case, the WPO disable distance is used to stop these calculations for objects that are far from the camera. This can significantly improve performance because the visual impact of the WPO effects diminishes with distance, making it unnecessary to compute them for distant objects. Next is an example of how to set this up.

With the static mesh spawner selected, on the right-hand side, search for **World Position Offset Disable Distance**. Set the value from 0 to 1000 units. This means that the WPO will work normally between 0 and 1000 units of its distance from the camera actor. This means that your character will observe the effects occurring within a range of 0 to 1,000 units/centimeters from the camera.

Figure 10.23 – Setting the World Position Offset Disable Distance value from 0 to 1000

In *Figure 10.24*, we see the WPO being activated or deactivated based on the distance between the camera's actor position and the assets that generate the WPO.

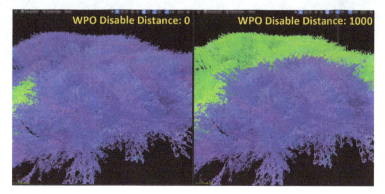

Figure 10.24 – WPO Disable Distance under different values

In the next section, we will explore other optimization techniques that will help to enhance the PCG setup even further.

Distance Field Lighting

Another use of optimization is to literally boost the framerate is to disable the specific lighting that only affects objects in the far distance. Search for **Affect Distance Field Lighting** and disable it in your **Settings** panel.

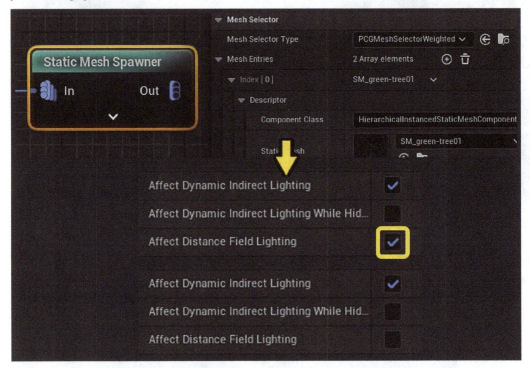

Figure 10.25 – Disabling Affect Distance Field Lighting

This will cause the shadows of the specified asset to become softer and diminish its internal quality, as we see in the following figure.

Figure 10.26 – Lower-quality shadows appearing inside the foliage

So far, we have learned how to optimize the PCG workflow using static mesh spawner settings. You can now test these changes in your project with extensive PCG-based foliage. In the following section, we'll explore how to optimize the textures within the texture settings so that they can keep the same quality without losing any compression.

Texture optimization

Proper texture sizes can significantly impact the overall performance of gameplay in PCG environments. It's crucial to optimize texture resolution for the meshes being spawned and distributed across the landscape.

For this example, we'll demonstrate how to reduce the size of a texture by adjusting its settings. Here's how you can do it:

1.   Locate the texture you want to optimize in the content browser and double-click to open it. In this case, I am going to use the textures of one of the leaves from the `EuropeanBeechTrees` folder.

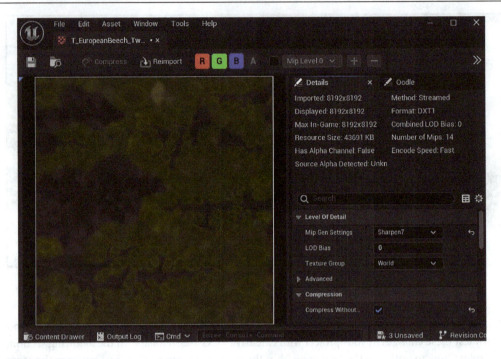

Figure 10.27 – Texture details panel

2.   The **Level Of Detail** texture settings help improve performance by using lower resolution textures when the object is far away from the camera. Under the **Level Of Detail** tab, change **Mip Gen Settings** from **Sharpen7** to **FromTextureGroup** to reset the texture to its default settings. Then, adjust the **LOD Bias** setting from 0 to 2. This will lower the texture quality, enhancing material performance.

Figure 10.28 – Changing the Level of Detail settings

3. Set a maximum texture size that is appropriate for the use case. For example, if the texture is used for distant objects, you can reduce the maximum size to save memory. By carefully managing texture sizes and settings, you can maintain high performance while still achieving good visual quality in your PCG environments. Under the **Compression** tab, navigate down to **Maximum Texture Size** and change its value from **8192** down to **1024**.

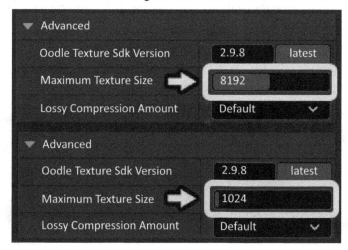

Figure 10.29 – Decreasing a resolution for the texture size

After changing the resolution in the settings, you can see the difference on the info panel. This will show how the resolution of the texture has drastically changed. Debugging inside PCG provides a visual output and this method helps to identify how each node can perform a specific task before the main static mesh is drawn on the landscape.

Figure 10.30 – Comparison of Details view – before and after changes applied

In the following section, we will delve into the debugging tools and their robust workflows. Additionally, we will examine how to test and debug each node for testing purposes.

Exploring debugging tools

This process has already been mentioned in the previous chapters of this book. However, it will be great to discuss it in more detail and express more interest in this topic. Debugging inside PCG provides a visual output and this method helps to identify how each node can perform a specific task before the main static mesh is drawn on the landscape. For this reason, let's present a PCG Graph example with a different nodes structure :

1. Within the graph, select the **Surface Sampler** node and press *D* to enable debug mode. This will display the current activity of the PCG graph when only the **Surface Sampler** node is functioning with the **Get Landscape Data node** .

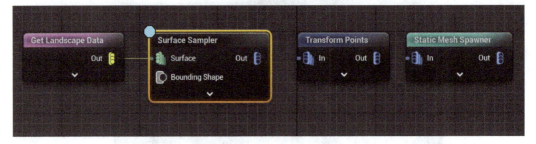

Figure 10.31 – Debugging Surface Sampler

When we test the debugging tool for **Surface Sampler**, small cubes will appear on the landscape surface, as shown in the following figure.

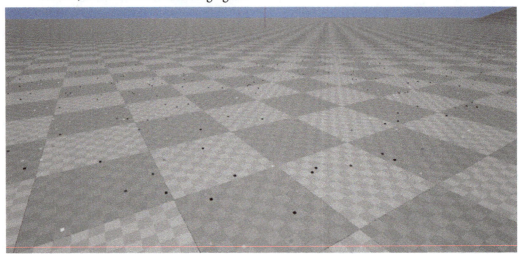

Figure 10.32 – Debugging visuals appearing in the form of smaller cubes

2. Let's use **Scale Method** on the right-hand side **Debug** panel to increase the cube sizes within the PCG volume. First, change its **Scale Method** setting from **Extent** to either **Absolute** or **Relative**. Then, adjust the **Point Scale** value to modify the size of the cubes. In my case, I have set the **Point Scale** value to 0 . 2 units.

Figure 10.33 – Debug cubes scaled up by a factor of 0.2

After changing the scale method to **Absolute**, we were able to adjust the scale values, making the cubes appear larger on the landscape surface.

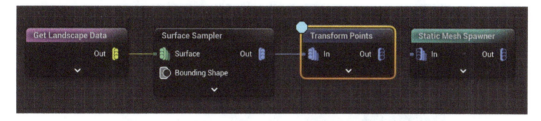

Figure 10.34 – The Transform Points node inDebug mode

3. Following *Figure 10.34*, let's try connecting the **Transform Points** node to the output of **Surface Sampler**. Make sure to disable the debug mode on **Surface Sampler** and enable it on the **Transform Points** node.

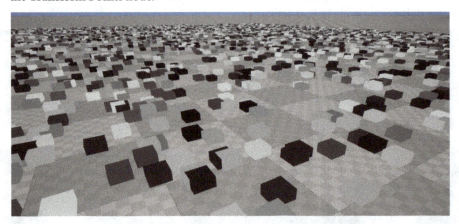

Figure 10.35 – Transform Points debug mode enabled

4. This time, we will change a few of the **Transform Points** settings. Set the scale to 5.0 for both **Min Scale** and **Max Scale**. Configure the rotation to −180 degrees for **Rotation Min** and 180 degrees for **Rotation Max**. This will ensure that the cubes rotate randomly, creating a randomized effect across the PCG volume!

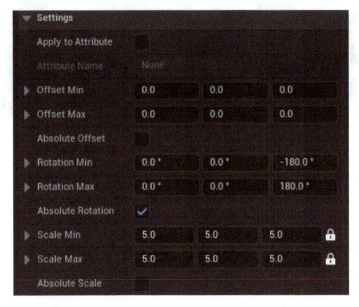

Figure 10.36 – Settings for the Transform Points node

After entering the numerical values for **Rotation Min** and **Rotation Max**, as well as **Scale Min** and **Scale Max**, we should see the effects applied to the scene.

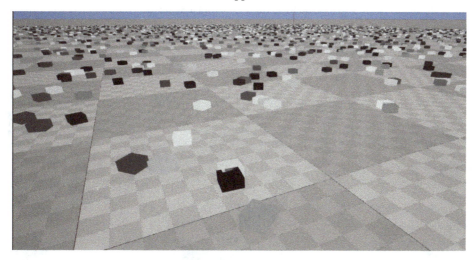

Figure 10.37 – Scale and the rotation values being applied to the Transform Points node

5. Last but not least, let's look at **Static Mesh Spawner**, which we will connect and test the PCG graph with a grass model. For this example, I used the PCG_Seedling_01 grass foliage model from the Plugins folder.

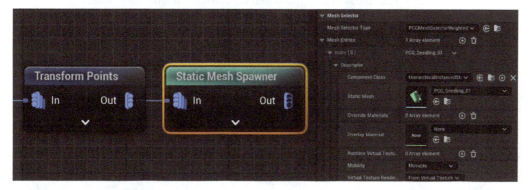

Figure 10.38 – Applying the PCG_Seedling_01 mesh to the Static Mesh Spawner mesh entry

6. As you noticed, the grass model doesn't follow the **Transform Points** scale, which is why it appears oversized. To adjust the grass model size, go to the **Transform Points** node and change the scale values from 5 to a **Scale Min** value of 1 and a **Scale Max** value of 2 (*Figure 10.40*).

Figure 10.39 – PCG_Seedling_01 applied with the different transformation values

We can reference the values set for the **Transform Points** node to randomize the rotation and scale of the foliage.

Figure 10.40 – Transform Points values in the Settings panel

7. This way of testing the node's parameter changes can be very helpful with debug mode. It helps to determine how the final model will be spawned across the entire landscape with only those small tweaks.

We are generating randomized grass positions on the landscape, which can be used to adjust the spacing between each grass clump.

Figure 10.41 – Simple grass models are being spawned slightly away from each other

In the next part, we will be getting to know some performance tools that come with Unreal Engine and how they can also optimize PCG performance.

Using Performance tools

In this section, we will explain how some tools can help to examine the scene and provide analytical readings to the performance of the scene that is currently running. Let's test out a few of the most essential tools that will be handy for your projects while creating a massive environment with the PCG plugin!

CPU profiling

CPU profiling enables you to precisely identify where your PCG processes are consuming the most time. By detecting these bottlenecks, you can optimize specific parts of your PCG graph to enhance overall performance. To test this, type stat scenerendering in the console command bar at the bottom of the screen. This command will display the performance metrics and draw calls currently active at your level.

Figure 10.42 – stat scenerendering cvar variable

The illustration here presents the stat menu report, which generates the performance results for the project's level. This stat setup provides comprehensive information about the current conditions as the CPU and GPU process all draw calls and post-process effects in real time.

Figure 10.43 – CPU profiling being enabled

If you want to find out more about the different types of profiling, especially for the CPU and GPU, it is worth looking into Unreal's documentation at `https://dev.epicgames.com/documentation/en-us/unreal-engine/testing-and-optimizing-your-content`.

Choosing scalability settings

This section introduces an optional addition to your projects that can enhance performance, depending on the platform you're targeting when using the PCG framework. In most cases, especially when developing games, it's common practice to stick with **Cinematic** or **Epic** quality to maintain high-level graphics. However, exceptions may arise when working on large-scale projects that require further optimization, particularly for mobile or virtual reality platforms. To address performance bottlenecks, adjusting the scalability settings is a widely used technique to improve graphical performance, especially when working within the level editor. This method reduces the overall quality of objects being rendered in the scene, helping to manage and optimize performance during editing, particularly for shadows, global illumination, reflections, and post-processing effects. Here's how it works:

1. With the main editor open, click on the **Settings** button on the right-hand side of the panel. This will open the **Settings** window menu. Hover down to **Engine Scalability Settings** with your mouse. Choose a **Medium** column.

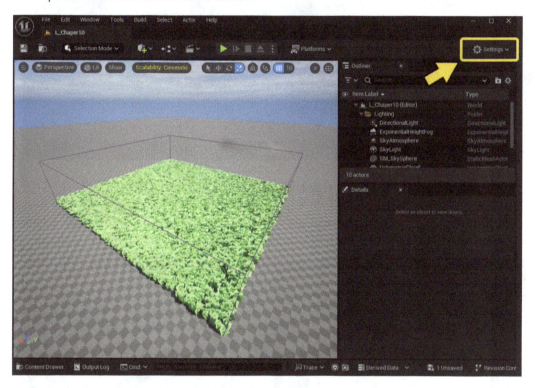

Figure 10.44 – Selecting the Settings option above the Outliner tab

The **Settings** menu will appear, and you need to select the **Engine Scalability Settings** menu to choose the appropriate optimization settings specifically for PCG use.

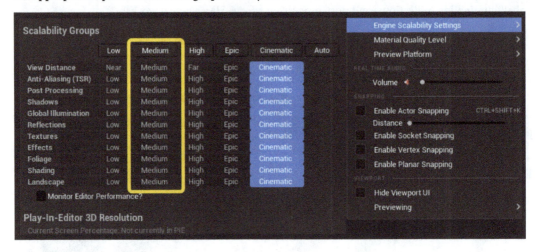

Figure 10.45 – Choose the most optimal quality option, which is the Medium level

You might notice some changes in the overall quality of the scene, but this approach is definitely helpful for addressing performance issues that may arise when working extensively with the PCG plugin!

Figure 10.46 – The difference between the Cinematic and Medium levels

You have just completed *Chapter 10* and learned how to effectively optimize your project using PCG. Additionally, you gained valuable insights into other optimization tools that help enhance performance further. The optimization methods discussed in *Chapter 10* will provide valuable lessons for any projects you work on using the PCG framework.

Summary

We have finally reached the end of this chapter, where we learned how to further optimize our PCG plugin. This has been a valuable experience in reducing the number of draw calls you might encounter during your world-building projects. PCG can present a heavy processing load, making it crucial to avoid potential issues that could crash your project, as we learned in this chapter. By applying these optimization tips, you can mitigate such risks in your future development.

In the next chapter, we will explore tips and guidelines for adding different types of nodes to our PCG graph and how each node can enhance our PCG workflow.

Get This Book's PDF Version and Exclusive Extras

UNLOCK NOW

Scan the QR code (or go to `packtpub.com/unlock`). Search for this book by name, confirm the edition, and then follow the steps on the page.

Note: Keep your invoice handy. Purchases made directly from Packt don't require an invoice.

11

Cheat Sheets, Extra Tips, and Shortcuts

At last, we have arrived at the final chapter of the book, *Chapter 11*! In this chapter, we will explore some tips and valuable information about the nodes, utilizing this knowledge to effectively visualize their purpose on the PCG graph.

In this chapter, we will cover the different node formations of each PCG graph and see how you can recognize their function by using one of the visual examples.

We will also use sample meshes from the `Plugin` folder and will enable the Debug mode in most of the visual examples. This chapter will provide valuable insights into the right formulas to build different PCG structures. We will use PCG volume samples and spline samples to compare and explain their main applications in various scenarios. Volumes and splines are integral to constructing PCG graph structures in projects, particularly when determining the type of environment to create. These elements significantly influence the procedural generation process, allowing for precise control over the placement and behavior of assets within the scene. By examining their uses, we can better understand how to leverage these tools to achieve desired environmental effects and optimize performance.

In this chapter, we will delve deeper into the following:

- PCG volume samples
- PCG volume splines

Technical requirements

You will need the following hardware and software to complete this chapter:

- A good computer that can run a PCG. It will need a multi-core CPU (AMD Ryzen 7/9, Intel i7/i9), GPU (NVIDIA RTX, AMD Radeon RX with 8+GB VRAM), and at least 16 GB RAM.

- The Unreal Engine version used in this chapter is version UE 5.4. The PCG tool was introduced with Unreal Engine 5.3 but some nodes got deprecated in Unreal Engine 5.3, hence the latest version of Unreal Engine is the most beneficial for this chapter.

All the examples are laid out inside the project, which you can find in the GitHub repository, and you can download this project from this link: `https://github.com/PacktPublishing/Procedural-Content-Generation-with-Unreal-Engine-5`.

The code in action video for the chapter can be found at `https://packt.link/JxEOL`

PCG volume samples

In this section, I will present several practical case examples as part of the cheat sheets and shortcuts. These examples demonstrate various uses of PCG graphs within PCG volumes, visually represented for easy understanding. You can reproduce these examples in your personal projects, saving you time and effort in searching for solutions online.

In this section, we will focus on PCG graphs primarily centered on the use of PCG volumes, showcasing how these samples can be beneficial for your projects.

Surface Sampler

The Surface Sampler node's primary function is to generate a set of points on the surface of a landscape or a specified geometry. These points can then be used as positions for spawning procedural elements in your scene. The following figure shows the Surface Sampler node in action:

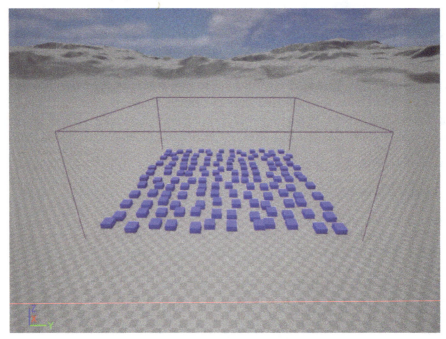

Figure 11.1 – The use of Surface Sampler

In order to activate the Surface Sampler node's ability to generate results on the landscape, we need to add another pin under the **Settings** panel, which you can find on the right-hand side after selecting the input node. Add another index array, change its **Allowed Types** value to **Surface**, and rename it Landscape. The Surface landscape will be viewed as a green input pin:

Figure 11.2 – Setting up the Input node to work with the Landscape Actor

Here is the snippet of the PCG graph code example running in Debug mode:

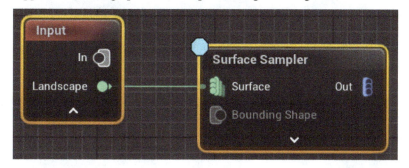

Figure 11.3 – Surface Sampler node under Debug Mode

Volume Sampler

The primary purpose of the Volume Sampler node is to generate a set of points within a given 3D volume. These points can then be used to create volumetric effects or to populate a volume with procedural elements. The following figure shows the use of the Volume Sampler node:

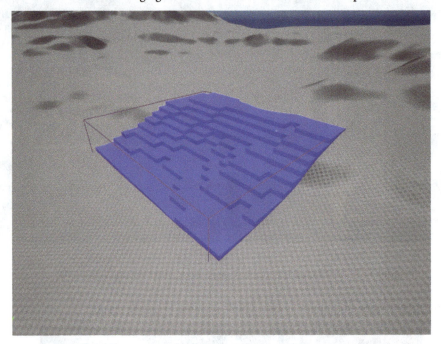

Figure 11.4 – Volume Sampler

Here is the snippet of the PCG graph code example running in Debug mode:

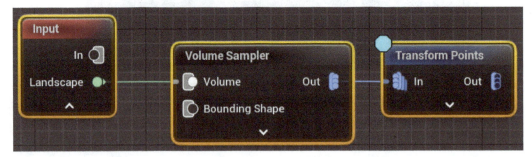

Figure 11.5 – The Volume Sampler node working together with the Transform Points node

Create Points Grid

The Create Points Grid node is designed to generate a grid of points within a specified volume or area. This node is particularly useful for creating structured and evenly spaced arrangements of points, which can be used for various PCG tasks such as placing objects in a grid pattern, defining sampling points for terrain, or setting up a regular layout for architectural elements. The following figure shows the Create Point Grid node in action:

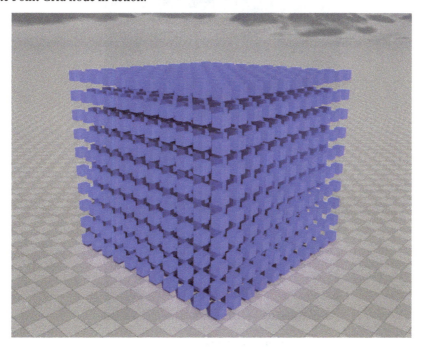

Figure 11.6 – The Create Point Grid node

Here is the snippet of the PCG graph code example running in Debug mode:

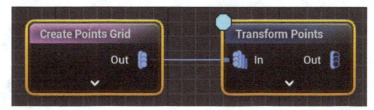

Figure 11.7 – The Create Point Grid node working with the Transform Points node

Get Texture Data

The Get Texture Data node is a utility node that extracts data from a texture and uses that data to influence the procedural generation process. This node is particularly useful for driving procedural content based on texture information, such as heightmaps, masks, or color gradients. The following figure shows the results of the Get Texture Data node and how it works together with the Texture sample:

Figure 11.8 – The Get Texture Data node

Here is the setup of the Get Texture Data node, including the Texture sample:

Figure 11.9 – The Get Texture Data node settings together with the Texture sample

Sample Texture

The Sample Texture node is a powerful tool for extracting and utilizing information from textures to drive procedural generation. This node allows you to sample various types of data from a texture, such as color, height, or alpha values, and use that data to influence the placement, transformation, or attributes of procedural elements. The following figure represents the results of the Sample Texture node working together with the Get Texture Data node:

Figure 11.10 – The Sample Texture node working together with the Get Texture Data node

Here is the complete PCG graph formula for spawning the static meshes based on the texture information extracted from the Get Texture Data node:

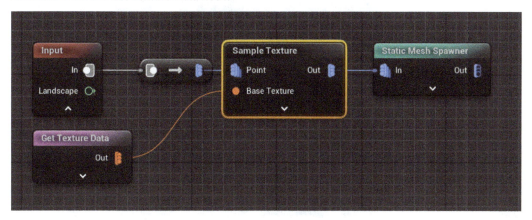

Figure 11.11 – The Sample Texture node working together with the Get Texture Data node

Density Remap

The Density Remap node is used to adjust the density values of points based on a defined remapping function. This node allows for more fine-tuned control over how density values are interpreted and applied, enabling the creation of more sophisticated and nuanced procedural content distributions.

In *Figure 11.12*, we have enabled Debug mode on the Density Remap node, allowing it to work in conjunction with the Surface Sampler node:

Figure 11.12 – The use of the Density Remap node

Figure 11.13 shows an example of how the Density Remap node can be utilized with two different settings, each featuring distinct remap distribution strengths based on their respective minimum and maximum input and output ranges:

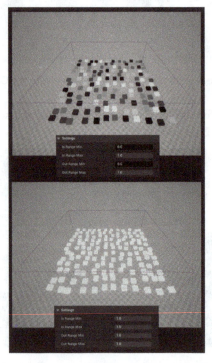

Figure 11.13 – The Density Remap node under two different configuration settings

Density Filter

The Density Filter node is used to selectively filter points based on their density values. This node helps refine the procedural generation process by removing points that do not meet specified density criteria, allowing for more precise control over where procedural elements are placed.

Here is the formula example for the PCG graph:

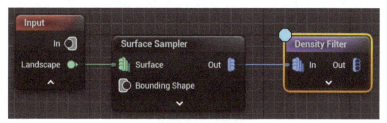

Figure 11.14 – The use of the Density Filter node

In the following figure, you will find two different configurations for each Density Filter example based on the **Lower** and **Upper Bound** values:

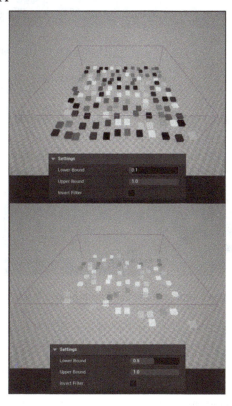

Figure 11.15 – Two different configurations for the Density Filter node

Self Pruning

The Self Pruning node is used to optimize and refine the distribution of procedural points by removing points that are too close to each other based on a specified distance threshold. This process helps to prevent the overcrowding of procedural elements and ensures a more even and realistic distribution. The following figure represents the use of the Self Pruning node together with the spawned static meshes:

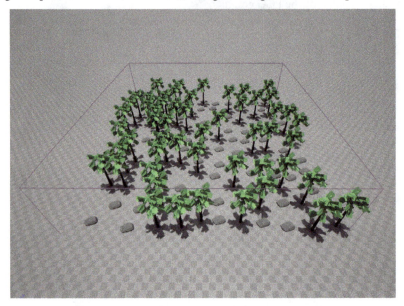

Figure 11.16 – The use of the Self Pruning node

Here is the snippet of the PCG graph code example of the Self Pruning use case:

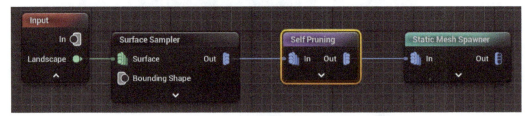

Figure 11.17 – The use example of the Self Pruning node inside the PCG graph

Bounds Modifier

The Bounds Modifier node is used to adjust the spatial boundaries within which procedural points are generated or modified. This can be particularly useful for defining specific areas in which certain procedural rules apply, allowing for more targeted and efficient content generation

When used together, the Bounds Modifier and Self Pruning nodes allow you to not only define the spatial limits of your procedural content but also ensure that the content within those limits is optimally distributed.

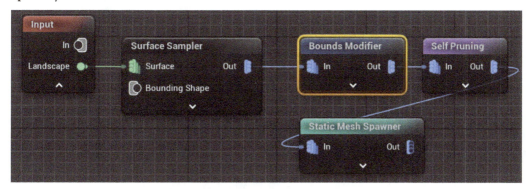

Figure 11.18 – The use of the Bounds Modifier node combined with the Self Pruning node

Here is an example of two different configurations and how they can differ depending on their **Bounds Min** and **Bounds Max** values:

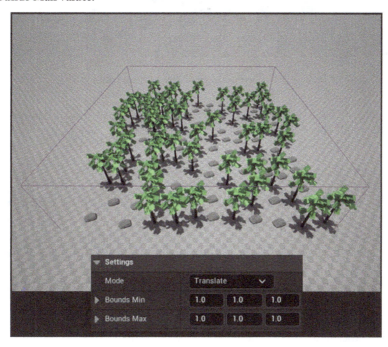

Figure 11.19 – An example of the Bounds Modifier node settings

The following figure shows a different configuration of the Bounds Modifier node in action:

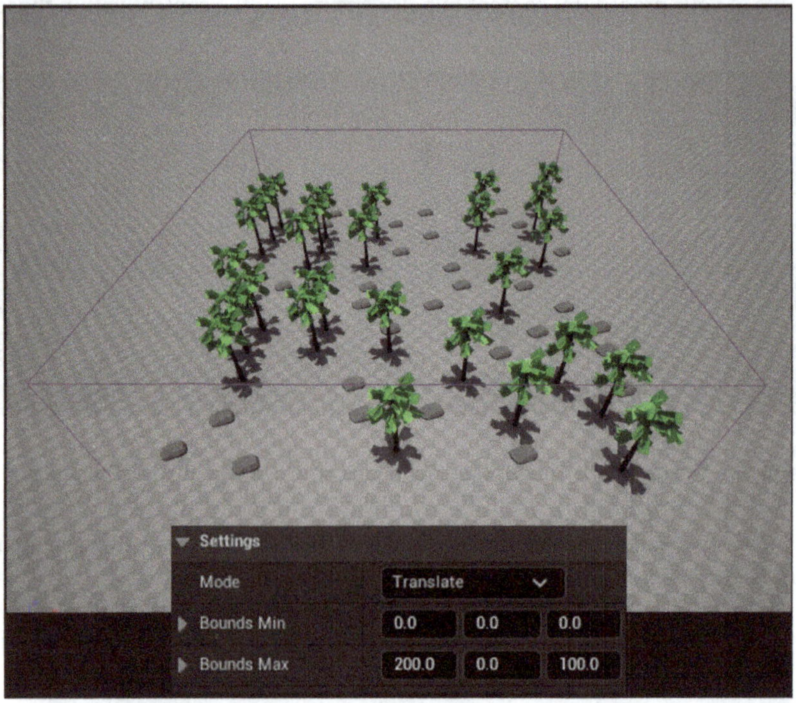

Figure 11.20 – The Bounds Modifier settings working

In the upcoming section, we'll dig into additional examples of PCG use cases, focusing on the Actor Blueprint with a Spline component to control the arrangement of PCG volumes.

PCG volume splines

In this section, we will explain how you can use the PCG graph in a way that correlates to the use cases of the spline controller via the Actor Blueprint. You can find the spline controller actor blueprint inside the GitHub project of *Chapter 11*. The following list presents examples of spline Component usage across various PCG node configurations for the spline.

Spline Sampler (On Interior)

The Spline Sampler node is used to sample points along and inside the spline. This node is particularly useful for creating procedural content that follows a specific path or shape defined by a spline, such as roads, rivers, environments, or other linear and closed-loop features.

The primary purpose of the Spline Sampler node is to generate points along a spline and inside the interior spline closed loop. These points can then be used to place procedural elements in a way that follows the path of the spline, ensuring that the generated content adheres to the intended layout and design. In the following figure, we use the **On Interior** setting for **Dimension** to generate results within the closed loop of the **Spline** component:

Figure 11.21 – Setting up the On Interior Dimension settings

The next figure showcases the results of the spawned cubes, which are positioned inside the closed loop on the scene:

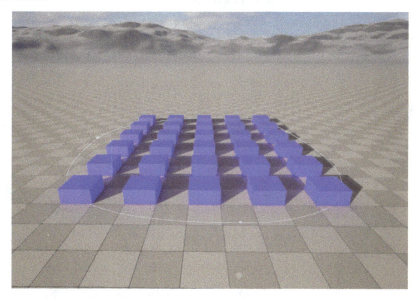

Figure 11.22 – Spawned cubes within the closed loop spline

Projection

The Projection node is used to project points onto a surface or geometry. This is particularly useful for aligning procedural points with the contours of a target surface, such as projecting points onto a terrain, mesh, or other 3D objects.

The primary purpose of the Projection node is to take a set of points and project them onto a target surface or geometry. This ensures that the points conform to the shape and contours of the target, making it ideal for tasks such as placing objects on uneven terrain or aligning elements with complex surfaces. The following figure demonstrates the Projection node connected with the other nodes in action:

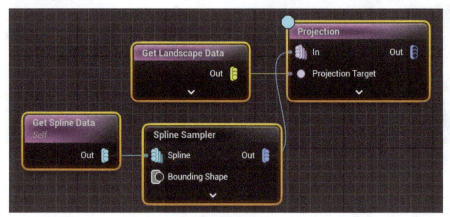

Figure 11.23 – An example use of the Projection node

The results are as follows, showcasing the cubes that are evenly spawned across the Landscape surface:

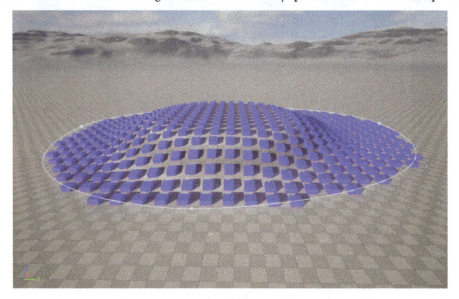

Figure 11.24 – The Projection node in Debug mode visualized on the scene

Spline Sampler (On Spline)

As we discussed previously, the Spline Sampler node plays a major role in creating a path for the spawned objects to follow. This time, we are using the **On Spline** settings (see *Figure 11.25*) to follow the curve spline path. The spline path will procedurally generate cubes along the spline within the specified distance range.

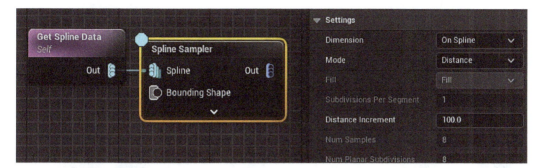

Figure 11.25 – Changing the Dimension setting to On Spline

The following figure demonstrates the cubes generated along the spline curve:

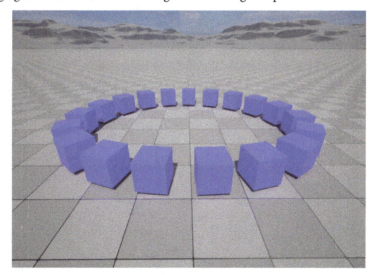

Figure 11.26 – Cubes spawned across the spline path

Difference

The Difference node is used to subtract one set of points from another, effectively filtering out points that overlap or intersect between two point sets. This operation is particularly useful for creating more complex and refined procedural distributions by removing unwanted points from a primary set based on their presence in a secondary set.

The primary purpose of the Difference node is to compare two sets of points and retain only those points from the primary set that do not overlap or intersect with the points from the secondary set. This helps in creating more precise procedural content by excluding points that should not be present based on specific criteria.

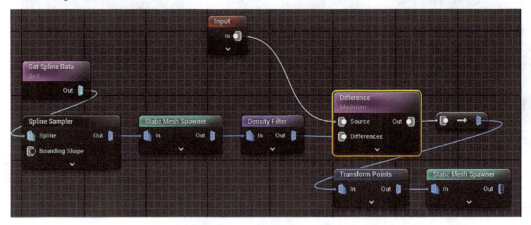

Figure 11.27 – The use of the Difference node inside the PCG graph

The following figure accurately represents the use of the Difference node inside the level:

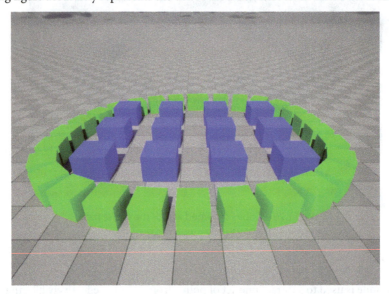

Figure 11.28 – A good use case of the Difference node with the spline controller

Copy Points

The Copy Points node is used to duplicate points from one set to another while optionally applying transformations or modifications to the copied points. This node is particularly useful for creating variations or multiple instances of procedural elements based on an initial set of points.

The purpose use of the Copy Points node is to create duplicates of points from an input set, allowing for the replication of procedural elements while, optionally, applying transformations such as translation, rotation, or scaling.

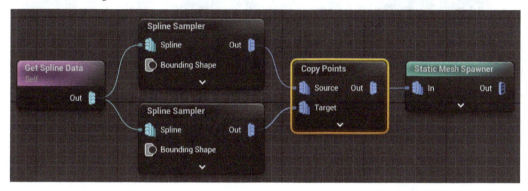

Figure 11.29 – Good demonstration case of the Copy Points node in PCG Graph

The following figure demonstrates a good example of using Copy Points with the node structure in *Figure 11.29* inside the PCG graph:

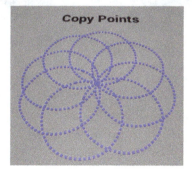

Figure 11.30 – A good use case of Copy Points nodes inside the level

Spawn Actor

The Spawn Actor node is used to instantiate actors in the game world based on procedural point data. This node is particularly useful for dynamically placing actors, such as characters, props, or interactive elements, according to the logic defined within the PCG graph.

The primary purpose of the Spawn Actor node is to create instances of specified Actor classes at the locations and with the attributes defined by procedural points. This allows for the dynamic and automated placement of actors in the game world, driven by procedural generation logic.

Here is the snippet example of the PCG graph node structure using a Spawn Actor node:

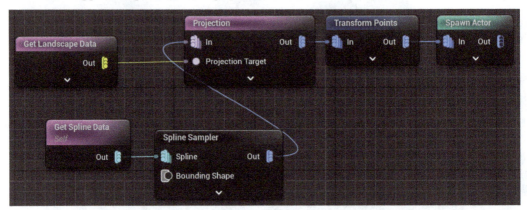

Figure 11.31 – The use of the Spawn Actor node with the other nodes

The following figure shows the results of the Spawn Actor node using the basic Actor Blueprint of a chair:

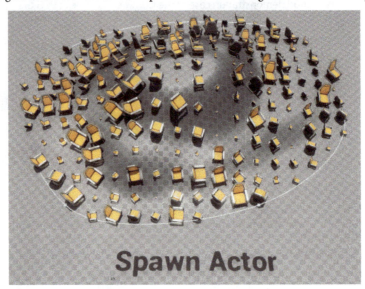

Figure 11.32 – The Actor Blueprint of the chairs being spawned across the Landscape

Mesh Sampler and Set Mesh As Attribute

Here's another example of using a mesh sampler that is generated for each vertex point on every mesh within the PCG volume. What's really cool is that you can use the vertex or triangle locations to generate an additional set of meshes on top of the existing geometry:

Figure 11.33 – The Set Mesh As Attribute node generates chairs on top of the original static meshes

For this reason, I'd like to introduce one last trick for the PCG Framework. I'll show you how I achieved this effect and guide you through the steps. Let's dive in:

1. In this exercise, we will use two meshes included with the project. The first is a custom-made geometry, **SM_Mesh**, which can be found in Content | PCG | GeometryScript | Meshes. The second is a chair model, **SM_Chair**, from the **StarterContent** pack, located under Content | StarterContent | Props:

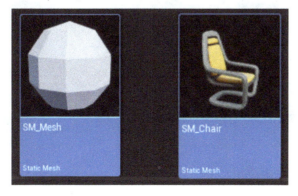

Figure 11.34 – Static Meshes Overview

2. To work with the Mesh Set Attribute Spawner, you must first enable **Procedural Content Generation (PCG) Framework Geometry Script Interop** in the **Edit | Plugins** section. When prompted with a message indicating that this is an experimental feature, click **Yes** and then select the **Restart Now** button to reboot the project:

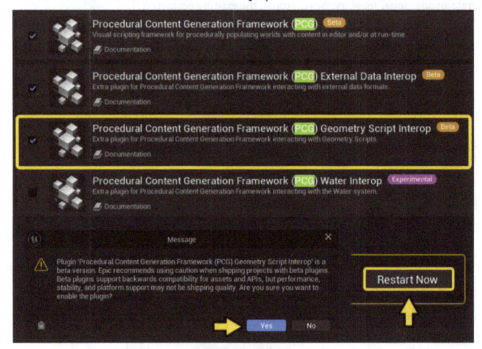

Figure 11.35 – Enabling Procedural Content Generation Framework
(PCG) Geometry Script Interop inside the Plugins window

3. Inside your Content | PCG | GeometryScript folder, create two **PCG Graph** assets. Let's name them PCG_MeshSamplerGenerator and xPCG_MeshSpawner:

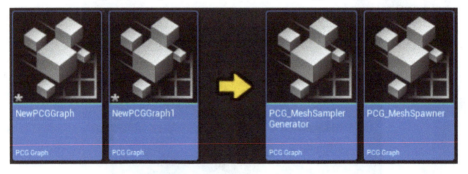

Figure 11.36 – Creating two new PCG Graphs and naming them PCG_
MeshSamplerGenerator and PCG_MeshSpawner

4. In the same folder, right-click, and under the **Blueprint** tab, search for the **Structure** asset. Rename your **Structure** asset S_Meshes:

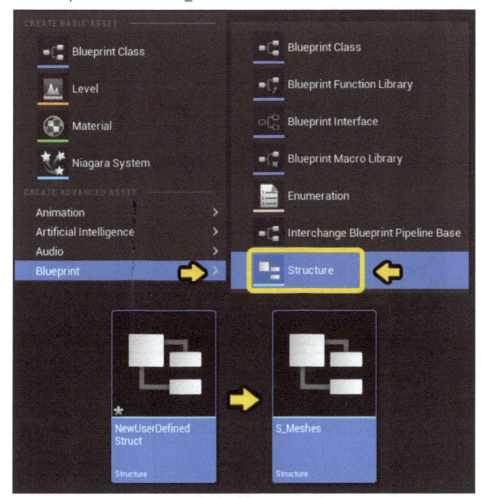

Figure 11.37 – Adding the Structure asset from the Blueprint menu

5. Double-click the **S_Meshes** asset. Inside the asset, click **+Add Variable** twice to create two variables. Set the first variable to **Static Mesh** and rename it `Mesh`. Set the second variable to a **Float** type and rename it `Weight`:

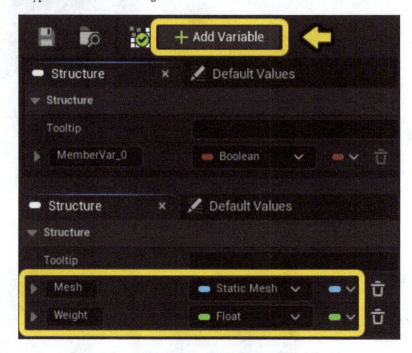

Figure 11.38 – Adding two variables for the Mesh Static Mesh and the float variable for Weight

6. Let's open the **PCG_MeshSpawner** graph. Inside the PCG graph editor, select the **Input** node and go to the **Settings** panel on the right. Change the **Usage** value to **Loop** and **Allowed Types** to the **Point** type:

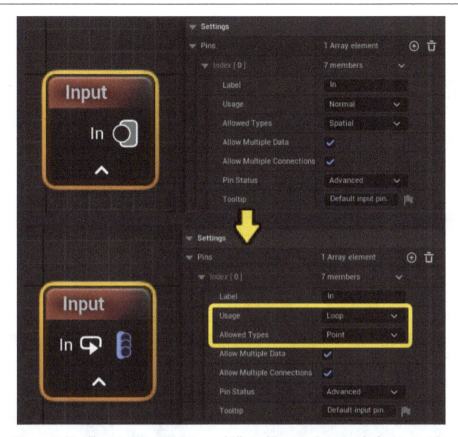

Figure 11.39 – Changing Usage to Loop and Allowed Types to Point inside the Input node

7.   The next step is to add all the following nodes, which we will use to spawn the meshes together with the **Attribute** mesh from **Point Index**. Add the following nodes: **Get Attribute From Point Index**, **Transform Points**, **Mesh Sampler**, and **Copy Points**.

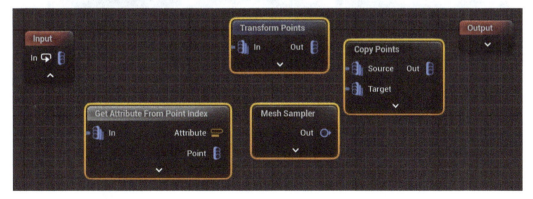

Figure 11.40 – Adding all the nodes inside the PCG_MeshSpawner PCG graph

8. To begin with, let's set up the **Get Attribute From Point Index** node and rename **Input Source** to Mesh:

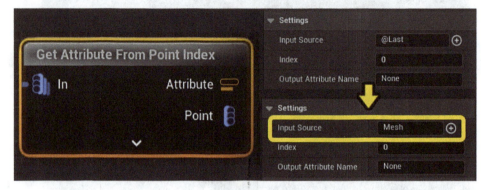

Figure 11.41 – Renaming Input Source to Mesh inside the Get Attribute From Point Index node

9. Open the **Mesh Sampler** node and connect the **Attribute** output of the **Get Attribute From Point Index** node to the **Static Mesh** input of the **Mesh Sampler** node:

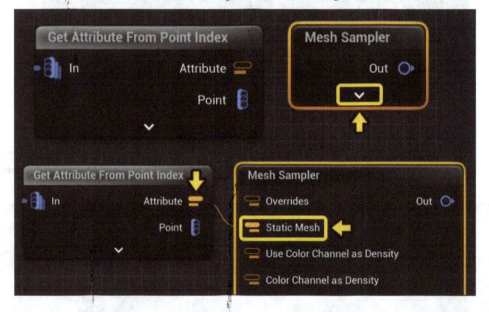

Figure 11.42 – Connecting the Attribute output to the Static Mesh input

10. Select the **Transform Points** node and set **Rotation Max** for the **Z** axis to 90 degrees. Then, adjust **Scale Min** to 0.5 and **Scale Max** to 0.8 for all axes:

Figure 11.43 – Changing the Rotation Max, Scale Min, and Scale Max values inside the Transform Points node

11. Select the **Output** node and go to the **Settings** tab on the right. Change **Allowed Types** to **Point**:

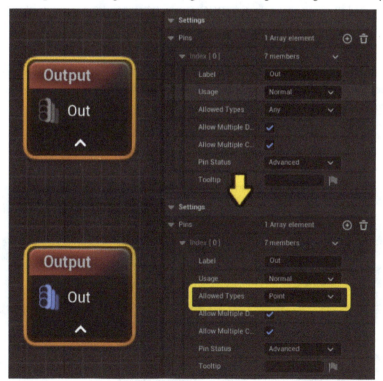

Figure 11.44 – Changing Allowed Types to the Point type inside the Output node

12. Connect all the nodes by following the number sequence exactly as shown in the following figure. Lastly, connect the **Input** node directly to the **Target** output of the **Copy Points** node:

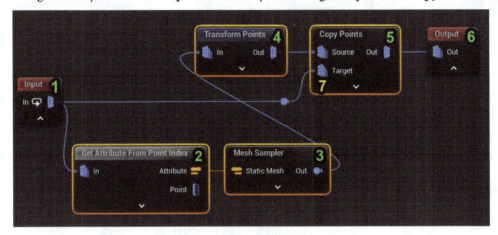

Figure 11.45 – Connecting all the nodes together by following the numbered sequence

We've now completed the creation of the first PCG graph, **PCG_MeshSpawner**. In the next section, we'll focus on building the **PCG_MeshSamplerGenerator** graph. Here, we will construct a node structure that works with **PCG_MeshSpawner** as a subgraph.

I'll guide you on how to incorporate the **PCG_MeshSpawner** graph into **PCG_MeshSamplerGenerator** and use the other nodes to generate meshes from the subgraph by using the attributes that match the right meshes from the content folder. Let's dive in!

1. Open the **PCG_MeshSamplerGenerator** graph and add the following nodes into the PCG graph editor: **Get Landscape Data**, **Surface Sampler**, **Match And Set Attributes**, **Attribute Partition**, **Add Attribute**, and **Static Mesh Spawner**.

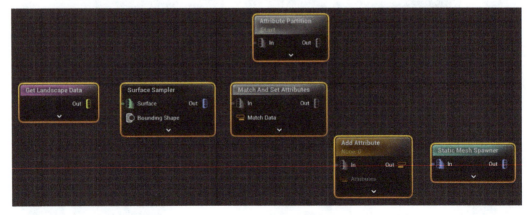

Figure 11.46 – Adding all the necessary nodes inside the PCG_MeshSamplerGenerator PCG graph

2. Select the **Surface Sampler** node and, in the **Settings** tab on the right, set the **Point Extents** values for both **X** and **Y** to 200.0 units. This adjustment will space out the newly generated meshes from each other.

Figure 11.47 – Changing the Point Extents values for the Surface Sampler node

3. Select the **Match And Set Attributes** node and, in the **Settings** tab on the right, enable **Match Weight Attribute** and change the name to Weight. This will allow us to access the **Weight** variable from the **S_Meshes** asset.

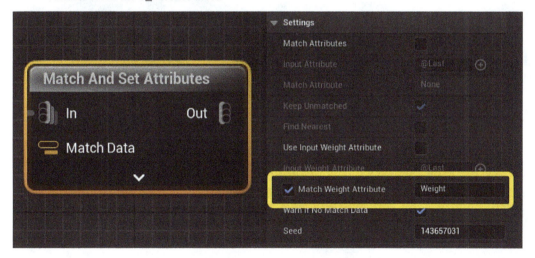

Figure 11.48 – Enabling the Match Weight Attribute setting and renaming the attribute to Weight

4. Select the **Attribute Partition** node and, in the **Settings** tab on the right, change the **Index[0]** array name to Mesh. This will allow us to access any mesh assets that are called **Mesh** from the **S_Meshes** asset.

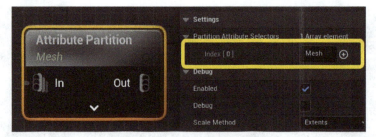

Figure 11.49 – Renaming the attribute to Mesh under the Attribute Partition node

5. Next, select the **Add Attribute** node and, in the **Settings** tab on the right, rename **Output Target** to Mesh. Then, change the **Type** value to **Soft Object Path**, and finally, assign the **SM_Chair** model to **Soft Object Path Value**:

Figure 11.50 – Changing the settings for the Add Attribute to access the SM_Chair static mesh

6. Select the **Static Mesh Spawner** node and, in the **Settings** tab, go to the **Mesh Selector** tab. Set **Mesh Selector Type** to **PCGMeshSelectorByAttribute** and change **Attribute Name** to Mesh:

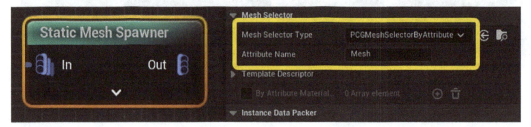

Figure 11.51 – Changing Mesh Selector Type to the PCGMeshSelectorByAttribute
type of the Static Mesh Spawner node

7. For the final step, add one more node to the PCG graph. Right-click in the PCG graph editor, search for SubGraph Loop, select it, and add the **Loop** node to the graph.

Figure 11.52 – Adding the SubGraph Loop node inside the PCG graph

8. Select the **Loop** node and, on the right under the **Instance** tab, search for and add the **PCG_MeshSpawner** PCG graph:

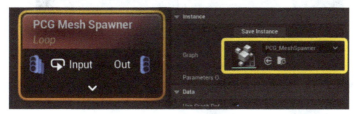

Figure 11.53 – Adding the PCG_MeshSpawner PCG graph instance for the SubGraph Loop node

9. With all this setup done, let's connect all the nodes together by following the numbers exactly as presented in the following figure:

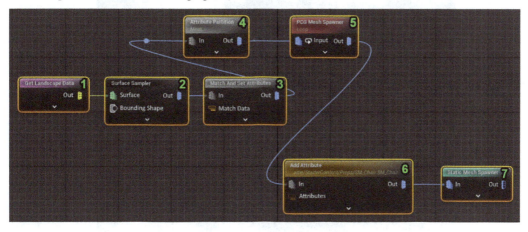

Figure 11.54 – Connecting all the nodes together by following the numbered sequence

10. The next step is to add another variable. At the top of your PCG graph, click on the **Graph Settings** button and go to the **Instance** tab. Click the + sign next to the **Instance** tab to add a new property, then search for the S_Meshes variable. Finally, add the **SM_Mesh** static mesh to the **Mesh** array:

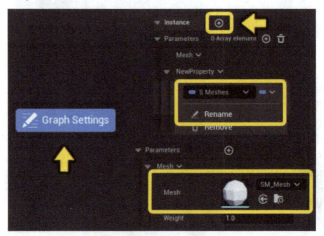

Figure 11.55 – Adding a new Instance variable that links to the S_Meshes asset and including the SM_Mesh static mesh in the Mesh array

11. Next to the **Match And Set Attributes** node, right-click on the graph and search for the **Get Mesh** variable. Connect the **Mesh** variable to the **Match Data** input of the **Match And Set Attributes** node:

Figure 11.56 – Adding a Mesh variable to the graph and connecting to the Match Data input of the Match And Set Attributes node

12. Lastly, connect the output of the **Match And Set Attributes** node to the **Static Mesh Spawner** node. This will provide direct control over the spawned static meshes.

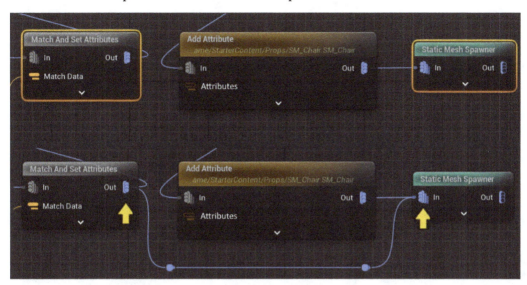

Figure 11.57 – Connecting the Match And Set Attributes node directly to the Static Mesh Spawner node

13. The following figure shows the final structure of the nodes:

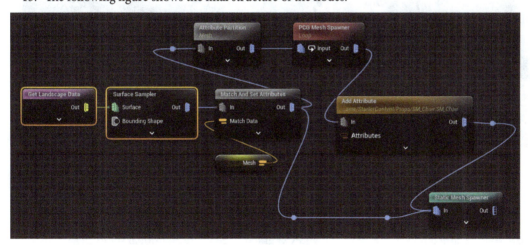

Figure 11.58 – The final PCG_MeshSamplerGenerator PCG graph

Now, you can drag and drop **PCG_MeshSamplerGenerator** into the scene and test the PCG graph:

Figure 11.59 – Adding PCG_MeshSamplerGenerator to the level to view the final results

That wraps up *Chapter 11*! I hope you enjoyed exploring the new features of the PCG Framework. By now, you've learned how to create your own PCG graph to generate additional static meshes on top of those produced by the initial PCG graph.

Summary

Congratulations on reaching the final chapter of the book! In this chapter, we have gained valuable insights into the additional functionalities of each node and how they produce different results with the right PCG graph configurations. These techniques will be incredibly useful for both your personal and professional projects.

Remember, the key to mastering procedural content generation lies in experimentation and creativity. Don't be afraid to try new configurations and push the boundaries of what's possible. The examples provided are just starting points—your imagination is the only limit!

Whether you're crafting detailed game environments, simulating natural landscapes, or designing complex architectural scenes, always approach each project thoughtfully. Pay attention to the details, and continuously refine your techniques to create stunning and immersive PCG environments using the powerful and effective PCG plugin.

Thank you for embarking on this journey with me. I hope this book has equipped you with the knowledge and inspiration to create amazing things. Happy creating, and best of luck in all your future endeavors!

Index

U

V

W

packtpub.com

Subscribe to our online digital library for full access to over 7,000 books and videos, as well as industry leading tools to help you plan your personal development and advance your career. For more information, please visit our website.

Why subscribe?

- Spend less time learning and more time coding with practical eBooks and Videos from over 4,000 industry professionals

- Improve your learning with Skill Plans built especially for you

- Get a free eBook or video every month

- Fully searchable for easy access to vital information

- Copy and paste, print, and bookmark content

Did you know that Packt offers eBook versions of every book published, with PDF and ePub files available? You can upgrade to the eBook version at packtpub.com and as a print book customer, you are entitled to a discount on the eBook copy. Get in touch with us at customercare@packtpub.com for more details.

At www.packtpub.com, you can also read a collection of free technical articles, sign up for a range of free newsletters, and receive exclusive discounts and offers on Packt books and eBooks.

Other Books You May Enjoy

If you enjoyed this book, you may be interested in these other books by Packt:

Cinematic Photoreal Environments in Unreal Engine 5

Giovanni Visai

ISBN: 978-1-80324-411-2

- Generate a Master Material to create hundreds of different material instances
- Explore lighting principles and apply them to UE lighting systems
- Evaluate the pros and cons of real-time rendering in the world-building process
- Build massive landscapes with procedural materials, heightmap, landmass, and water
- Populate an environment with realistic assets using Foliage and Megascan
- Master the art of crafting stunning shots with Sequencer
- Enhance visual quality with Post Process Volume and Niagara
- Produce a photorealistic shot using the Movie Render Queue

Practical Game Design

Adam Kramarzewski, Ennio De Nucci

ISBN: 978-1-80324-515-7

- Define the scope and structure of a game project
- Conceptualize a game idea and present it to others
- Design gameplay systems and communicate them clearly and thoroughly
- Build and validate engaging game mechanics
- Design successful games as a service and prepare them for live operations
- Improve the quality of a game through playtesting and meticulous polishing

Packt is searching for authors like you

If you're interested in becoming an author for Packt, please visit `authors.packtpub.com` and apply today. We have worked with thousands of developers and tech professionals, just like you, to help them share their insight with the global tech community. You can make a general application, apply for a specific hot topic that we are recruiting an author for, or submit your own idea.

Share Your Thoughts

Now you've finished *Procedural Content Generation with Unreal Engine 5*, we'd love to hear your thoughts! Scan the QR code below to go straight to the Amazon review page for this book and share your feedback or leave a review on the site that you purchased it from.

https://packt.link/r/1801074461

Your review is important to us and the tech community and will help us make sure we're delivering excellent quality content.

www.ingramcontent.com/pod-product-compliance
Lightning Source LLC
Chambersburg PA
CBHW060642060326
40690CB00020B/4488